Knowledge Management in Construction

Edited by

**Chimay J. Anumba, Charles O. Egbu
and Patricia M. Carrillo**

Foreword by
Sir Michael Latham

 **Blackwell**
Publishing

Editorial offices:
Blackwell Publishing Ltd, 9600 Garsington Road, Oxford OX4 2DQ, UK
Tel: +44 (0) 1865 776868
Blackwell Publishing Inc., 350 Main Street, Malden, MA 02148-5020, USA
Tel: +1 781 388 8250
Blackwell Publishing Asia Pty Ltd, 550 Swanston Street, Carlton, Victoria 3053, Australia
Tel: +61 (0)3 8359 1011

First published 2005 by Blackwell Publishing Ltd

3   2007

ISBN: 978-1-4051-2972-5

Library of Congress Cataloging-in-Publication Data
Anumba, C. J. (Chimay J.)
    Knowledge management in construction/Chimay Anumba, Charles Egbu, and Patricia Carrillo — 1st ed.
        p. cm.
    Includes bibliographical references and index.
    ISBN: 1-4051-2972-7 (hardback : alk. paper)
    1. Construction industry — Information systems.   I. Egbu, Charles O.   II. Carrillo, Patricia.   III. Title.

TH215.A5865 2005
690 — dc22
2004020191

A catalogue record for this title is available from the British Library

Set in 9.5/12.5pt Palatino
by Techbooks Electronic Services Pvt Ltd
Printed and bound in Great Britain
by TJ International Ltd, Padstow, Cornwall

For further information on Blackwell Publishing, visit our website:
www.blackwellpublishing.com/construction

Contents

Foreword

The lessons learned on many construction projects are often lost when the project team is disbanded at the end of a project and the parties involved move on to new projects. This results in much reinventing of the wheel and repetition of past mistakes. The situation is compounded by the fact that there are few appropriate mechanisms for capturing and sharing the new knowledge gained on projects between all members of a project team. Post-project reviews go some way towards capturing lessons learned, but the haphazard manner in which these are usually conducted means that insufficient time is normally available, or participants have forgotten key aspects of learning events. Furthermore, very few companies willingly share their learning from projects for fear of exposing their mistakes.

In the light of the above, knowledge management is essential for improving the construction project delivery process. Fortunately, the importance of sharing knowledge and learning is now increasingly being recognised in the construction industry, with many organisations appointing knowledge managers or others with the remit to harness and nurture the organisation's knowledge. This may work well at organisational level, but there is still considerable scope for better collaborative learning at project level by all project participants. Effective knowledge management at project level is best undertaken within the context of the collaborative working practices that I advocated in my report on the construction industry *Constructing the Team*. The Egan Report, *Rethinking Construction*, reinforced the principles outlined in my report and, amongst other things, recommended the establishment of a movement for change as a means of sustaining improvement and sharing learning.

This book builds on both my report and Sir John Egan's by seeking to introduce the key elements of knowledge management to the construction industry. The authors cover a wide range of issues – from basic definitions and fundamental concepts, to the role of information technology and engendering a knowledge-sharing culture. Practical examples from construction and other industry sectors are used throughout to illustrate the various dimensions of knowledge management. The challenges in implementing knowledge management are outlined and the ensuing benefits highlighted.

It is essential that appropriate knowledge management processes are put in place if the construction industry is to continuously improve its business processes and sustain the productivity and other improvements that have followed the publication of both my report and Egan's. This book provides practical guidance on how this can be done and I consider it essential reading for all participants in the construction process.

Sir Michael Latham

Preface

A considerable proportion of the rework, delays, mistakes and cost overruns on construction projects can be attributed to poor knowledge management. While many organisations have some elements of knowledge management practice, which are not necessarily labelled as such, there is much more that can be done to improve the construction project delivery process through better management of the knowledge generated on projects and in individual firms. There are serious dangers for companies that ignore knowledge management – they run the risk of simply repeating past mistakes or worse, taking decisions that can lead to major disasters. On the other hand, organisations that proactively manage their knowledge stand to reap considerable rewards (in cost savings, process efficiencies, reductions in errors and rework, etc.) and will be able to deliver more innovative solutions to their clients.

The importance of knowledge management (KM) is increasingly being recognised in the construction industry, with leading organisations beginning to formulate strategies and appointing Chief Knowledge Officers or Knowledge Managers to deliver the strategy. This is laudable but it is important that the implementation of knowledge management initiatives is not based simply on a desire to 'join the bandwagon' but on a deep-seated conviction of the positive impact that effective knowledge management can make on construction business processes. A detailed understanding of the role of knowledge management in construction organisations is therefore essential.

This book is intended to introduce the subject of knowledge management to practitioners and researchers in the construction industry. It contains both the theoretical background to knowledge management and the practical insights, applications and methodologies necessary to enable organisations to implement KM. It draws from a large body of research work and industry practice and is essential reading for all sectors of the construction industry, particularly those businesses which seek to rise above the competition.

Chimay J. Anumba
Charles O. Egbu
Patricia M. Carrillo

Acknowledgements

We are grateful to all contributors who have spent a considerable amount of time putting together their chapters. We acknowledge the contribution of the various agencies and organisations that funded the research projects and initiatives on which this book is based. Dr Herbert Robinson (one of the contributors) played a major role in collating the chapters. We are also indebted to our families, whose continued love and support make ventures of this nature both possible and worthwhile.

Contributors

Dr Ahmed M. Al-Ghassani is the Assistant Dean at Salalah College of Technology in Oman. Prior to that, he was Head of the Construction Department and has also served for several years in the private and public sectors. His research interests include improving the process of structural design, knowledge management and IT use in construction. He obtained his PhD from Loughborough University, UK, an MSc in Structural Engineering from UMIST, UK and a BSc from SQU, Oman.

Professor Chimay J. Anumba holds the Chair in Construction Engineering and Informatics and is founding Director of the Centre for Innovative Construction Engineering at Loughborough University, UK. He has recently been a Visiting Professor at the Massachusetts Institute of Technology (MIT) and Stanford University, USA. His research work cuts across several fields, including knowledge management, and has received support with a total value of over $13 million from a variety of sources. He has contributed over 250 publications in these fields. His industrial experience spans over 20 years and he is a Chartered Civil/Structural Engineer, holding Fellowships of the Institution of Civil Engineers the Institution of Structural Engineers and the Chartered Institute of Building.

Dr Patricia M. Carrillo is a Senior Lecturer in Construction Management in the Department of Civil and Building Engineering at Loughborough University, UK, and Visiting Professor at the University of Calgary, Canada, and the University of Colorado at Boulder, USA. She has investigated companies' knowledge management practices since 1999. Her experience stems from six different knowledge management research projects involving UK consultants and contractors, leading Canadian oil and gas companies and Canadian engineering, procurement and construction (EPC) companies. She is a Chartered Civil Engineer and a Fellow of the Chartered Institute of Building.

Dr Peter Demian has recently completed his PhD on knowledge management and design knowledge reuse at the Project Based Learning Laboratory (PBL Lab), in the Civil and Environmental Engineering Department at Stanford University. His research interests include design theory, human–computer interaction and information retrieval. He holds a master's degree from Stanford University, and bachelor's and master's degrees from Cambridge University.

Professor Charles O. Egbu holds the Chair of Construction and Project Management at Glasgow Caledonian University, UK. He has received substantial grants from major UK research funding agencies and has published over 100 articles, with many relating to innovation and knowledge management in construction. He worked in industry before obtaining his PhD from the University of Salford, UK. He is a visiting Fellow at the Management, Knowledge and Innovation Research Unit (Open University Business School), and a member of several knowledge management and innovation groups of the International Council for Research and Innovation in Building and Construction (CIB). He is a corporate member of the Chartered Institute of Building (CIOB) and the Association for Project Management (APM).

Dr Patrick S.W. Fong is Associate Professor in the Department of Building and Real Estate at Hong Kong Polytechnic University. He is Adjunct Professor in the Faculty of Design, Architecture and Building at the University of Technology, Sydney. He has published articles and book chapters focusing on managing knowledge in projects, teams and professional services firms. He is currently Vice President of the Hong Kong Knowledge Management Society.

Dr Renate Fruchter is the Founding Director of the Project Based Learning Laboratory (PBL Lab), lecturer in the Department of Civil and Environmental Engineering, and thrust leader of 'Collaboration Technologies' at the Center for Integrated Facilities Engineering (CIFE), at Stanford University. She leads a research group focusing on the development of collaboration technologies for multidisciplinary, geographically distributed teamwork and e-learning. Her research interests focus on R&D and larger-scale deployment of collaboration technologies that include web-based team building, project memory, corporate memory, interactive workspaces for exploration and decision support, and mobile solutions for global teamwork and e-learning.

Dr John M. Kamara is a lecturer and coordinator of the Architectural Informatics Group at the School of Architecture, Planning and Landscape, University of Newcastle upon Tyne. He has over 60 scientific publications in the fields of project development (client requirements processing), collaborative design and construction, knowledge management and informatics. He is Principal Investigator on an EPSRC/industry-funded project on the Capture and Reuse of Project Knowledge in Construction (CAPRIKON).

Dr Ian Lyttle is Knowledge Management Competency Leader in IBM Business Consulting Services, UK. In his consulting practice, he focuses on aligning technical and social dimensions of knowledge to organisational goals. He has been involved with the development of IBM's Intellectual Capital Management System since 1993 and has addressed numerous conferences on this theme. He has published *Knowledge as a Corporate Asset: A Case Study of a Knowledge Network*.

Dr Dominique Poole studied Architecture at Oxford Brookes University and received a PhD for her thesis on technical innovation from the client's perspective. Her research, carried out with Buro Happold, provides an understanding of issues that influence client decisions regarding innovative practices and solutions, and the implications of these for the designer and others involved in the construction process. She has experience of project-based learning within the construction industry and has analysed the effectiveness of communities of practice. She has particular expertise in the fields of technical innovation and knowledge management, which is currently being applied at Arup Research and Development.

Professor Paul Quintas holds the Chair of Knowledge Management at the Open University Business School, UK. He has been researching, teaching and advising in the area of the management of knowledge and innovation for about 20 years. His work has achieved international recognition through his publications and also his work as an adviser to the OECD, the UN, four Directorates of the European Commission and to national governments.

Dr Herbert S. Robinson is a Senior Lecturer at London South Bank University. He lectures in knowledge management to undergraduates and post-graduates, including engineering doctorate students, and has published articles on knowledge management and business performance. After graduating from Reading University, he worked with consultants Arup (UK) and on World-Bank-funded projects in The Gambia, before returning to academia to pursue his research interests. He holds a PhD in Systems and Infrastructure Management.

Tony Sheehan is Group Knowledge Manager and an Associate Director of Arup, where he has developed and implemented the global knowledge management strategy. Tony has balanced expertise in the areas of cultural change, process management and knowledge management technology. He has presented at numerous knowledge management conferences, universities and business schools, and has published articles in such publications as *Intranet Strategist*, *Knowledge Management Review* and *Sloan Management Review*. Tony has an MBA in Engineering Management from Loughborough University, a first degree in Materials Science from Oxford University, and is a Chartered Engineer.

Carys E. Siemieniuch is Senior Lecturer in Systems Engineering at Loughborough University, UK. A systems ergonomist for 17 years, with both UK professional and European CREE registration, she has expertise across the full range of systems-related human factors topics including knowledge lifecycle management (KLM) systems and organisational and cultural aspects of enterprise modelling techniques.

Murray A. Sinclair is a Senior Lecturer in the Department of Human Sciences at Loughborough University, UK. Current research areas concern advanced manufacturing technology, computer-supported co-operative working in the extended enterprise, knowledge lifecycle management and processes for managing continuous change and capability acquisition. He is a Fellow and Council Member of the Ergonomics Society, a registered European Ergonomist and member of the IEEE Engineering Management Society.

1 Introduction

Charles O. Egbu, Chimay J. Anumba
and Patricia M. Carrillo

This book is about knowledge and knowledge management (KM) from a construction industry perspective. Its principal aims are to provide practitioners, academics, students and those interested in the construction industry with an improved awareness and understanding of KM principles and practice. It is also intended to provide guidance on the use of appropriate strategies, techniques and technologies for exploiting knowledge, for the benefits of the industry, projects and individual 'knowledge workers'.

Although knowledge is an age-old subject, one that occupied Plato and Aristotle and many other philosophers before them, the last two decades have witnessed an explosive growth in discussions about knowledge – knowledge work, knowledge management, knowledge-intensive organisations and the knowledge economy – albeit mainly in a non-construction industry context. The main impetus for the growth in KM in the last two decades includes globalisation and increased competition, diffusion of new information and communication technologies, financial implications of intellectual property rights, newer procurement routes, new work patterns, employment rights and contracts, and also contradictory political and ethical underpinnings. An important aspect of this book, therefore, is to explain this new emphasis on an enduring subject.

This book brings together work from key, selected contributors that offers insights, raises awareness and provides advice on managing knowledge. The contributors adopt varying forms of critical and reflective approaches on the knowledge discussions, informed by their empirical, theoretical and practical experiences on the subject.

In Chapter 2, we are reminded that the concept of KM is not new, that it has been with us for many centuries. However, the chapter presents a strong case for the prominence and rise of KM in recent decades. The chapter also succinctly presents the nature of knowledge dimensions, including the distinctions between tacit and explicit knowledge. It truly expounds on the complexities of knowledge, its social dimensions, its 'stickiness' and its context specificity. An appreciation of knowledge sharing across organisational boundaries and the importance of absorptive capacity when reflecting on how to learn from external sources are documented. An understanding of the complexities of knowledge allows us to understand the challenges in requisite processes needed for managing it in dynamic and complex situations.

In Chapter 3, a case is made for managing knowledge in a project-based environment. The construction industry, like many project-based industries, has its own peculiarities which impact upon KM. These include having to work in projects with start and finish dates, where 'short-termism' is rife, where different professionals and workers are brought together for a relatively short period and then disperse. Construction workers on site, in the main, are contracted on a short-term basis and live a more-or-less nomadic way of life, moving from one project to another. Harnessing and consolidating certain aspects of knowledge assets for organisational use is difficult in these situations. In many ways, the construction industry is still seen as a 'traditional' industry. However, things are changing. The construction industry is also, more often than not, easily associated with its tangible products such as buildings, roads and other structures. Viewing the industry as a 'service' provider often seems to take a back seat. Yet, the industry is made up of different professionals and tradesmen and -women who offer their services and knowledge in the work they do. This chapter informs us that given the nature of construction markets today, the greater demands from clients, the increasing level of awareness of clients, the complexities of projects in terms of size, cost, technology and the increasing nature of collaborative forms by which clients procure construction works, the construction industry is increasingly becoming a more knowledge-intensive industry. A case is made for the need to value knowledge workers in the construction industry, as these are the people who provide important skills and knowledge in the provision of services for the industry in what is a knowledge economy.

Chapter 4 considers two very important aspects of the KM debate. First, it considers the strategies for KM in the context of the construction industry. In doing this, it cleverly addresses what KM means to construction. Second, the chapter addresses the important issue of the need for, and the challenges of, a business case for KM.

The chapter argues that for construction organisations, good KM practice requires knowledgeable people who are supported by integrated information and data sources in order to generate informed decision making. It also presents some KM strategies that have been found to be successful within construction organisations. Readers would benefit from the examples of successful strategies drawn from organisations such as Arup, Taylor Woodrow, Thames Water and Mott McDonald. It is noted that organisations could seek to capture construction knowledge in documents, databases or intranets. This 'explicit' knowledge approach often works well for standardised problems, but does little to enable the exchange of new ideas. Alternatively, organisations could focus primarily on people in order to develop ways in which they can exchange their 'tacit' knowledge to facilitate innovation. This approach, however, allows excessive duplication and wheel reinvention in some cases. It is argued that an appropriate balance of 'explicit' versus 'tacit' approaches depends on each organisation's strategy and the particular case in point. An organisation is bound to require elements of both approaches, and must integrate the two effectively.

Attention is also given to the role of people, processes, culture and technology as important embodiments of a robust strategy for KM. For example, we are informed that having an open and participative culture which values the skills and contributions of employees at all levels is critical to a successful KM initiative. In terms of successful processes in larger organisations, the chapter suggests embedding good KM practices into normal organisational processes to help transfer knowledge across the business. Creating a network of virtual knowledge leaders is useful in cascading good KM practices. Developing a network of divisional knowledge activists to help 'spread the word' is equally useful. It is also beneficial to align KM with quality assurance (QA) following the ISO 2000 approach such that KM can become embedded in the organisation. For a project-based industry such as construction, the chapter argues that project learning is complementary to an overall KM strategy. Post-project (or after action) reviews and learning histories are two important aspects of project learning.

Although much is said and written about different aspects of KM, not as much it seems is discussed about making a case for KM. The point, however, is made that since KM can represent a significant investment for an organisation, it is reasonable to apply it to business areas that will yield the best and most value.

We are informed in this chapter that a business case should be seen as a critical component of the KM initiative. The business case can be used to communicate the KM initiative to others, to establish a method for performance measurement and for receiving funding approval for the initiative. A good business case should be able to help answer some key questions often posed during the early stages of deciding to set up a KM initiative or project. Such questions include: Why is the organisation or unit embarking on the KM initiative? What is the initiative about? What is the likely cost of the initiative? What are the likely benefits or return on investment? How will overall success and failure of the initiative be measured?

This chapter also reflects on the 'value of knowledge'. In doing so, we are informed of the notion of assessing and quantifying the impact of knowledge initiatives or projects through such 'soft' approaches as communities of practice, quality circles and story telling. Similarly, issues associated with making a business case for KM, which considers quantifying intangible assets, are addressed.

Chapter 5 reflects on organisational readiness for KM. The premise on which this chapter is based and the message it attempts to present can be best summed up in the following words: 'If you would plant roses in the desert, first make sure the ground is wet'. This chapter introduces an important concept, which is *knowledge lifecycle management* (KLM). The point of this is that organisational knowledge could be said to have a lifecycle; it is discovered, captured, utilised and, eventually, retired (or lost) rather than killed. It is argued that if one views organisations as 'knowledge engines' providing value to customers, then the processes of KLM are vital for organisational survival. This line of enquiry helps us to comprehend organisations as acquiring and using capabilities in the form of knowledge and skills in knowledge-intensive processes to deliver value-for-money products. In addition, the consideration of organisations as 'knowledge engines' helps, in part, to explain the need to focus attention on the prime, essential assets of the organisations; on strategic considerations; and understanding that the organisation's knowledge rather than its physical assets primarily differentiates it from its competitors.

Preparing the organisational context for KLM is all-encompassing. It has human, organisational and technological dimensions. In addition, there are strategic and tactical issues to be considered. The chapter informs us that as part of this, organisational design needs to reflect the firm's role in its supply chain. Knowledge workers (people) must be empowered and resourced to execute organisational activities. Other issues highlighted in the chapter as being important as part of organisational readiness for KM include the importance of building trust through leadership, identifying and populating 'knowledge evangelist roles', establishing 'ownership' policies for knowledge, identifying and implementing workable security policies, creating generic processes and procedures, providing robust technical infrastructure, establishing review procedures to ensure discussion of knowledge capture, the reviewing of reward policies and establishing personal performance measures for knowledge sharing.

Since organisations rarely take time to evaluate whether they are in fact in a position to implement policies and procedures to manage knowledge, the views and discussions presented in this chapter should be useful for organisations contemplating to do just that. Finally, the chapter reminds us that preparing an organisation for KLM is an interactive learning process which has positive and negative feedback loops, allowing for lessons and changes to be made during the process.

In Chapter 6, we are introduced to the different tools and techniques available for KM. The point, however, is made that the term 'knowledge management tools' should not be narrowly defined to mean just information technology (IT) tools. They also include non-IT tools required to support the sub-processes of KM. The chapter carefully distinguishes IT tools for knowledge management (KM technologies) and non-IT tools for knowledge management (KM techniques). A comparison between the two, together with examples, are provided in the chapter.

Organisations seem to encounter difficulties in identifying appropriate tools due to the range of competing products in the marketplace, overlap between the functions of various tools and the cost associated with acquiring and using them. The chapter examines existing approaches for selecting KM tools, discusses their limitations and goes on to present a new and innovative method as an alternative (SeLEKT – Searching and Locating Effective

Knowledge Tools). The method facilitates the selection of appropriate tools based on key dimensions of knowledge that reflect an organisation's business needs and context. It also considers, *inter alia*, whether the knowledge is internal or external to the organisation; 'ownership forms' reflecting whether knowledge is owned by individuals or shared; and the 'conversion types' reflecting the interaction between tacit and explicit knowledge.

This chapter also presents a comprehensive list of KM technologies and software applications classified by KM sub-processes and by KM 'technology families'.

In Chapter 7, we are introduced to cross-project KM. The central messages centre on the nature and characteristics of projects, the project environment (the context in which a project is initiated, implemented and realised) and how these impact upon how knowledge is managed in project-based environments such as construction. The chapter commences with a definition of a project, which is important in beginning to understand some of the subtle and fundamental differences between managing knowledge in a project-based environment as opposed to a non-project-based environment. Projects are temporary endeavours with start and finish dates, involve stated objectives and goals to be achieved, have resources that need to be expended, involve a set of interrelated activities, can often be one-offs (unique) and/or involve some elements of change.

The chapter makes the case that the changing nature of project organisations, for example, has implications on KM requirements across the lifecycle of a project. In this regard, we are reminded that each project lifecycle is made up of different activities carried out by different professionals and tradespeople, each having and also requiring different knowledge assets. The chapter also discusses the important issue of knowledge transfer across different projects, its dimensions in terms of the role of individuals, project reviews, contractual and organisational arrangements and the goal of knowledge transfer. This draws out the important issues of absorptive capacity of knowledge and how individuals and organisations learn, and how individuals in a supply chain are motivated to share and transfer knowledge. The concept of the 'knowledge dilemma' is brought to the fore. What knowledge should be shared and what knowledge should not in order to preserve the competitive advantage of an organisation. Similarly, it raises the issue of how knowledge transforms across the supply chains and the role of knowledge networks.

Another important contribution of this chapter focuses on the approaches for live capture and reuse of project knowledge, distinguishing between 'soft' concepts and 'hard' technologies. Soft concepts include collaborative learning and learning histories, whereas hard technologies include information and communication technologies such as project extranets, workflow management tools and other groupware applications for collaborative working. A conceptual framework for the live capture and reuse of project knowledge is described as a way forward for successful cross-project KM in the construction industry. This reflects a step change in the way construction project knowledge is managed. Project knowledge files, an integrated work-flow system and a project knowledge manager are important dimensions in the armoury needed to improve real-time knowledge capture.

Chapter 8 considers the role of KM as a driver for innovation. It argues that knowledge is highly associated with innovation. An organisation's capacity to innovate, it is suggested, depends to a very considerable extent upon the knowledge and expertise possessed by its staff.

The chapter also discusses the intrinsic and complex relationships between KM and innovation and the role of building and maintaining capabilities to facilitate the process. Since the innovation strategies of an organisation are constrained by their current position, and by specific oppor-tunities open to them in the future based on their competencies, construction organisations would need to determine their 'technological trajectories or paths'. This will involve taking due cognisance of the strategic alternatives available to them (e.g. organisational processes), their attractiveness, and opportunities and threats that lie ahead. The organisational processes adopted to integrate the transfer of knowledge and information across func-tional and divisional boundaries (strategic learning) are essential and need to be managed. It is equally argued that organisational core capabilities and core competencies, which are difficult to imitate or copy and which provide competitive advantage for innovation, are developed through a knowledge building process. The knowledge building process includes problem solving, future experimenting and prototyping, internal implementing and integrating and external importing of knowledge. Also, an important capa-bility is the expertise to manage internal and organisational complementary resources.

The chapter provides examples on how, through mobilising knowledge, experiences and technological skills, organisations can improve their ability to innovate: by focusing on a particular market niche, through novelty, through complexity, by stretching basic models of products and processes over a period of time, by continuous movement of the cost and performance frontiers, and by integrating personnel and the team around products and services.

Issues of strategy, process, structure, culture and technology and their impact on KM are also discussed. The implications for managers in man-aging knowledge for successful exploitation are discussed. It is suggested that managers have critical roles to play in making knowledge productive, in knowledge development and in the exploitation of knowledge for inno-vative performance. Specific ways of involving managers and deepening a manager's understanding of KM issues are also documented in the chapter.

In Chapter 9, attention is focused on performance measurement for KM initiatives and projects. We are reminded in this chapter that the most com-pelling argument for measuring the performance of knowledge assets and KM is to demonstrate its business benefits so that resources and support necessary for a successful implementation can be provided. It confirms that the old maxim 'you cannot manage what you cannot measure' also applies to KM. The chapter also presents examples of cost savings from KM programmes from some selected companies.

In considering types of performance measures, we are informed of two distinct aspects of KM that need to be measured. The first relates to knowledge assets (stocks) and the second to KM projects or initiatives (flow). Measures for knowledge assets seem to relate to what organisations know, and what they need to know or learn in order to improve performance. Measures of knowledge assets or stocks or intellectual capital focus on several components, including human, structural and customer capital. On the other hand, measures for KM projects or initiatives focus on the expected outputs of KM interventions. Cost-benefit, cost-utility and cost-effectiveness measures are often used in this regard.

The chapter also presents and makes a comparative assessment of the different performance measures for KM. These measures include the metric approach, economic approach and the market value approach. Other application tools such as business performance measurement models, Skandia Navigator, Intangible Asset Monitor, Human Resource Accounting, IMPaKT Assessor, KM Benefits Tree, Degussa–Huls Approach, Inclusive Value Methodology, market-to-book-value ratios, Tobin's q ratio and the Intellectual Capital Index are also discussed.

There are simple and complex performance measures. As organisations 'progress' to a stage where implementation of KM is mature and well co-ordinated, a more robust measure may be required.

In Chapter 10, the discussions centre on KM strategy development. Building on Chapter 4, which introduced the important issue of strategy, the chapter explores the concepts of 'supply-driven' and 'demand-driven' KM strategies, as well as 'mechanistic' and 'organic' KM strategies. While supply-driven KM initiatives assume that the fundamental problem of KM is to do with the flow of knowledge and information within organisations, the demand-driven approaches are more concerned with users' perspectives and their motivation and attitude. KM strategies can also be described as 'mechanistic' or 'organic' with respect to the emphasis on either 'explicit knowledge' (mechanistic) or 'tacit knowledge' (organic). Mechanistic approaches often rely on IT as compared to organic approaches (which focus on non-IT tools). In this chapter, we are also reminded of the importance of 'content' (the knowledge that is to be managed) and 'context' (organisational setting for the application of knowledge) in the success of KM strategies.

The chapter also addresses the '*how*' of KM in relation to the development of KM strategies within construction organisations by presenting and discussing a CLEVER (Cross-sectoral LEarning in Virtual EnteRprise) approach to KM. The CLEVER approach, which is linked to business drivers and goals, concentrates mainly on organisational and contextual aspects of

KM strategy in dealing with the 'definition' and 'analysis' of a knowledge problem in order to facilitate the selection of an appropriate strategy for KM in construction organisations.

Chapter 11 addresses the concept of corporate memory, focusing on design knowledge capture, sharing and reuse. Corporate memory is defined as a 'repository of knowledge in context'. Put simply, it is an 'external' knowledge repository containing the organisation's past projects that attempts to emulate the characteristics of an internal memory, i.e. rich, detailed and contextual. Corporate memory grows as the design firm works on more projects. In this chapter, we are also informed that the main reasons why knowledge reuse often fails, are that it is not captured, it is captured out of context, rendering it not reusable, or there are no formal mechanisms for finding and retrieving reusable knowledge. It also notes that reusing designs and design knowledge from an external repository of knowledge from previous projects fails due to the fact that state-of-practice archiving systems do not support the designer in finding reusable items and understanding these items in context in order to be able to reuse them. The chapter raises and attempts to answer very searching questions such as: What is the nature of knowledge capture, sharing and reuse? What are the key characteristics of the knowledge reuse process? How can the design knowledge reuse process in the architecture, engineering and construction (AEC) industry be supported by a computer system, and what are the natural idioms and interaction metaphors that can be modelled into a computer system to provide an effective knowledge reuse experience to a designer? The chapter presents and discusses innovative capture and interaction metaphors and their software implementation which act as a repository of knowledge in context. The presented prototype systems support knowledge capture, sharing, finding, reuse through project context exploration, and evolution history exploration in a large corporate memory that is made up of informal tacit and formal explicit knowledge.

In Chapter 12, attention is given to the building of a knowledge-sharing culture in construction project teams. Knowledge sharing, we are informed, relies on reaching a shared understanding of the underlying knowledge, in terms of not only 'content' but also the 'context' of knowledge. The chapter presents and discusses some findings from an industrial case study of knowledge sharing involving construction professionals and the client's specialist consultants. The importance of socialising, sharing positive as well as negative experiences, and knowledge 'shielding' to protect a client from its competitors are explored. Communication is seen as key in knowledge sharing. Competition appears to negatively influence knowledge sharing. It is also suggested that factors such as openness, motivation, trust and pressure of time impact upon knowledge sharing in complex ways.

This book concludes with a succinct recap of the main issues drawn from all the chapters. The implications of the issues raised in the various chapters for organisations, practitioners and for academia are also presented.

Managing knowledge effectively has the potential to provide substantial benefits to organisations, projects and individual workers. However, it is not easy. Having said that, there are certain steps to take to begin to put things in place, address challenges along the way and exploit potential benefits. These are the steps that the contributors to this book have attempted to provide. These steps need to take account of people, technology, structures, process, strategy, finance and the consideration of environmental, legislative and market conditions. KM should be seen as a continuous and long-term endeavour. It should also be seen as an asset and not a liability. That is the collective message of this book. If the book helps individuals and organisations in construction and other project-based industries, even in a very small way, towards improving their understanding of KM, then the primary aim of the book would have been realised.

2 The Nature and Dimensions of Knowledge Management

Paul Quintas

2.1 Introduction

Human activity is inconceivable without knowledge. The scope of knowing and types of knowledge are as wide and varied as all the varieties of human pursuits. It is palpably obvious that without creating, accumulating, sharing and applying knowledge, no human civilisation could have existed. Even though the phrase 'knowledge management' only came into common usage in the West during the last five years of the 20th century, it is emphatically *not* the case that the management of organisational knowledge processes began in the mid-1990s. The economic value of knowledge has been discussed for centuries, from the ancient Greeks to Adam Smith and Alfred Marshall, who in 1890 wrote: 'Capital consists in a great part of knowledge and organisation..... Knowledge is our most powerful engine of production' (Marshall, 1972). Even 50 years ago it was stating the obvious to note that knowledge provides the basis of whole industries, or that it plays a crucial role in the functioning of organisations, and indeed is the source of innovation and competitive advantage. There is little doubt that knowledge was indeed managed before 1995 – how else were the pyramids, the steam engine or the Apollo spacecraft built? This means that we have to acknowledge that the processes that create and apply knowledge in organisations are by no means always labelled 'knowledge management'. We need therefore to distinguish between formal or labelled 'knowledge management' and the unlabelled management of knowledge processes, which has a rather longer history.

Having established this prior managing of knowledge, we cannot ignore the fact that there has been a surge of interest in knowledge management (KM) in the West from the mid-1990s. This is even more evident in organisational *practice* than it is in the plethora of academic articles, books and conferences on the subject. Early 1990s KM initiatives in firms such as BP, Chevron, Shell, Hewlett Packard, Buckman Labs and Xerox, and the pioneering of intellectual capital reporting in Skandia (1994), all pre-date the academic KM publishing boom. In Section 2.2 we address the key question of what has driven this upsurge of interest in KM. Identifying these key drivers helps us understand the dimensions and dynamics and, arguably, the future trajectory of the phenomenon.

The 1990s interest in KM initially triggered the response that it was about doing things that were new. More recently many organisations have come to realise that it is also a process of *discovery*. That is, discovering knowledge processes that already happen, such as the sharing of knowledge amongst communities of practitioners, that were rarely if ever labelled 'knowledge management'.

The case studies of Honda, Matsushita and other firms in Nonaka and Takeuchi's influential and widely quoted book *The Knowledge-Creating Company* (1995) were not examples of designated KM initiatives but rather descriptions of actual knowledge processes of knowledge sharing, knowledge combination and so on. These were identified *post hoc* as examples of knowledge being managed.

In a sense, then, KM processes have been *discovered* rather than invented. Another example is story-telling, which has recently been discovered as being alive and well and providing a means of knowledge sharing in many organisations. Thus in scoping KM we have to take into account those knowledge processes that function and work well without the KM label. Conversely (and ironically) many so-called KM initiatives and tools that emerged in the late 1990s were less concerned with addressing real knowledge issues than these informal or existing processes that are not so labelled. In particular there has been a focus on codified knowledge – which limits the scope to no more than information management. This approach rather misses the point of introducing knowledge into the equation. The nature and scope of human knowledge is rather broader than that which can be encoded.

In addition to making us aware of pre-existing knowledge processes, perhaps the major benefit from raising the issue of knowledge is that it invites us to think differently about key organisational resources and processes. This means moving beyond the rather safer and certainly easier ground of data and information management. As Spender (1996) highlights, there is little point in introducing such a complex concept as knowledge into management thought and practice if we do not take seriously the characteristics of knowledge that make it special, and distinguishable from information.

An enhanced focus on knowledge provides opportunities for new thinking, both about and within organisations. It is particularly valuable to realise that there is more than one way to view knowledge. For example, perceptions of knowledge differ between cultures. Grossly simplifying a more complex geographical and epistemological variation, Western and Eastern traditions differ in their views of the extent to which knowledge can be separated from the knower. In the West, we tend to think about knowledge as a 'thing' or commodity that can easily be moved around, managed and traded. Conversely the Eastern traditions are more likely to emphasise the inseparability of what is known from the individual or groups that know it. Put another way, a non-Western approach would tend to emphasise *knowing* as a *process* rather than knowledge as a *thing*. Adoption of the latter approach may be more practical than it might at first appear, as this definition of KM from the Xerox Corporation illustrates (Cross, 1998, p. 11):

> 'Knowledge management is the discipline of creating a thriving work and learning environment that fosters the continuous creation, aggregation, use and re-use of both organisational and personal knowledge in the pursuit of new business value.'

The Xerox definition is strongly process and action oriented. It does not emphasise knowledge resources and assets, as many definitions and indeed initiatives do. Rather, it focuses on the processes of creating new knowledge and actively doing things with it.

So the enhanced focus on knowledge can benefit from a raised awareness of differences in cultural perceptions of knowledge and how it might be managed. Different views of knowledge are an opportunity for shifting our own mindsets, as well as helping us understand the other culture. A European survey of KM among 100 European business leaders (Murray and Myers, 1997) revealed some interesting cultural differences. In France, more than anywhere else in Europe, nearly a quarter of business leaders believed you *cannot* create any processes to help you manage knowledge. It is simply a matter of 'management ability'. In Germany, more than four out of five respondents considered their organisation to be already good at encouraging staff to share knowledge and to bring forward new ideas. In the UK the main KM focus was to exploit and control the knowledge that companies believe they already have. Most remarkably, almost a quarter of UK respondents said that *creating new knowledge* was *not* a key priority, compared with only 1% in Germany.

Following the Introduction, Section 2.2 addresses the key question of why knowledge management has come to prominence in the last decade. The answers to this question help us better understand not only what has driven the recent growth of knowledge management, but also what are the dimensions of the phenomenon that are likely to shape the agenda for KM in the future. Section 2.3 examines the nature of knowledge, highlighting key knowledge issues that are important and relevant, including the tacit (implicit) nature of human knowledge, the social nature and 'stickiness' or context specificity of knowledge. Section 2.4 discusses the importance of extra-organisational knowledge dimensions and absorptive capacity, reflecting the ability to learn from external sources. Section 2.5 outlines the key knowledge processes necessary for organisations to manage knowledge more effectively. These processes cut across all the themes or agenda or issues mentioned or discussed in previous sections. Section 2.6 concludes that knowledge processes have been managed for as long as organisations have existed, without the KM label. However, knowledge and the processes and challenges of managing it have become more dynamic and complex.

2.2   Why knowledge management *now* ? The drivers

The mid-1990s saw a surge of publications, conferences and consultant activity in the KM area. The publication of journal articles on the subject rapidly took off from 1996, as indicated in Figure 2.1. The first periodicals dedicated to the 'new' topic, including *Knowledge Management, Knowledge Inc., Knowledge Management Review* and the *Journal of Knowledge Management*, appeared in 1997. By 1997 the KM bandwagon was rolling. In 1996–97 there were over 30 conferences on KM in the USA and Europe, and an estimated US $1.5 billion KM consulting revenue was being spent annually. Many organisations created new posts with titles such as Chief Knowledge Officer, Knowledge Manager, Director of Intellectual Capital or similar.

As we observed in the Introduction, this current surge of interest follows a long history of rather lower-key attention to knowledge from economic and organisational perspectives. Partly this is a result of the complexity of

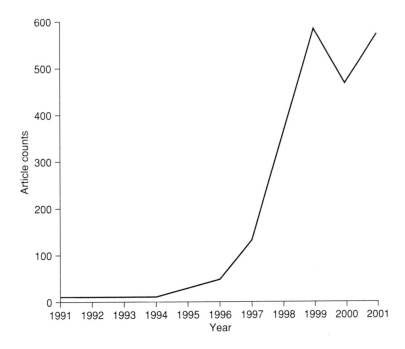

Figure 2.1   The number of knowledge management articles, 1991–2001. (Data reproduced with permission from Ponzi and Koenig 2002)

the subject, as Edith Penrose in her seminal *The Theory of the Growth of the Firm* in 1959 observed:

> 'Economists have, of course, always recognized the dominant role that increasing knowledge plays in economic processes but have, for the most part, found the whole subject of knowledge too slippery to handle.'

> (Penrose, 1959, p. 77)

The current idea of the knowledge economy also has its roots in Fritz Machlup's heroic attempt to map and measure *The Production and Distribution of Knowledge in the United States* (Machlup, 1962). Peter Drucker wrote in 1969 that knowledge had 'become the central capital, the cost centre and the crucial resource of the economy' (Drucker, 1969, p. ix). Scholars such as Daniel Bell built on the work of Machlup and others to propose the idea that the developed economies, having moved from agrarian to industrial in the 19th century, moved in the 20th century into a service- and knowledge-based *post-industrial* society. For Bell (1973), knowledge is the 'axial principle' of post-industrial society, and most of the current arguments for a knowledge-based economy were effectively put forward in his 1973 book *The Coming of Post-Industrial Society.*

The key question that emerges from the above is why did knowledge come to the top of the management agenda in the 1990s, a gap of some 100 years after Marshall (1972) recognised that 'Knowledge is our most powerful engine of production'? Our analysis suggests that a number of *drivers* have come together at this time. These fall under six headings, as we will see in the following section.

2.2.1   The drivers of knowledge management

1.   Wealth being demonstrably and increasingly generated from knowledge and intangible assets

Company value has come to be increasingly dependent on intangible assets, knowledge assets, intellectual capital and intellectual property. Perhaps the example that created the greatest impact in the early 1990s came when Microsoft, a relatively small company with less than 14 000 employees, was valued by the stock market (in terms of market capitalisation) to be worth more than IBM, which employed over 300 000 people and had an installed base of large computer systems across the world. What was the key to Microsoft's profitability and phenomenal success? They owned an intangible asset – the *de facto* standard for personal computer operating systems software, MS.DOS (and Windows). The market value of companies that own key intangible assets such as intellectual property rights, a standard or a brand may exceed the value of their conventional assets many times over. Roos and Roos (1997) state that in 1996, 94% of Microsoft's market value (US$119 billion) came from intangible assets. Similarly 85% of Intel's (US$113 billion) and 96% of Coca-Cola's (US$148 billion) market value were from intangible assets.

So, too, awareness of the value of knowledge and the ability to create it is evidenced in the large-scale investments in science-based and new-technology-based companies. Huge investments in biotechnology firms, many of which have no revenue income or show no profits, are based on the assumption that these firms will create knowledge that will lead to new product innovation and therefore competitive advantage. Previously public knowledge, such as that created in government laboratories and universities, became privatised, or the subject of proprietorial claims as patent lawyers moved into the laboratories (Massey *et al.*, 1992). Knowledge itself, and the capability to generate it, was drawn further under the aegis of private investment, ownership and competition.

The 1990s saw renewed attempts to measure and value intangible assets and knowledge. Accountants have long attempted to place a value on intangible assets under the headings 'goodwill' and brands, and also to value intellectual property, such as copyrights and patents. In the early 1990s a number of companies pioneered the development of intellectual capital auditing and reporting, with a view to capturing more accurate and comprehensive data on intellectual resources and capabilities. The Swedish financial services company Skandia developed an audit tool for intellectual capital – the Skandia Navigator – based on the Balanced Scorecard (Skandia, 1994). Other methods include the IC Index (Roos and Roos, 1997) and the Intangible Assets Monitor (Sveiby, 1997). Intellectual capital auditing and reporting is now used by many organisations internally as an input to strategy formulation and externally as a means of indicating organisational performance (e.g. Systematic, 2002).

2.   The rediscovery that people are the locus of much organisational knowledge

'If NASA wanted to go to the moon again, it would have to start from scratch, having lost not the data, but the human expertise that took it there last time.'

(Brown and Duguid, 2000, p. 122)

Ironically, certain negative aspects of human resource strategies in the 1980s and early 1990s have made their contribution to the current interest in managing knowledge. During this period in European and North American economies many organisations underwent programmes of 'downsizing' (e.g. by making employees redundant), 'de-layering' (e.g. by shedding layers of middle management) and 'outsourcing' (removing the capacity to do certain business functions, which then have to be bought in). One advocate of 'business process re-engineering' urged organisations 'Don't automate, obliterate!' (Hammer, 1990). The result of these strategies was that many organisations found they no longer retained the ability to react to change or even to fully understand their own markets and business. They had lost their corporate memory and capabilities they did not know they had or needed.

Following the downsizing, early redundancies and outsourcing of the 1980s, in the 1990s organisations rediscovered the central importance of people. In many cases staff that had been made redundant had to be rehired, often as consultants, because their knowledge was found to be irreplaceable. Once again people were recognised as possessing knowledge and know-how, having the ability to create knowledge and value, and collectively retaining organisational memory. Much of this individual knowledge is unknown to others and unmapped and unrecorded. The latter frustration is evident in much-quoted statements by the CEOs of Texas Instruments and Hewlett Packard:

> 'If TI only knew what TI knows' – Jerry Junkins, when CEO of Texas Instruments.

> 'I wish we knew what we know at HP' – Lew Platt, chairman of Hewlett Packard.

> (quoted in O'Dell and Grayson, 1998, p. 154)

The management of people, and the relationship between individuals, groups and organisational knowledge, are therefore central foci for KM. Often knowledge is created within communities of practice who share understandings and experience that are not easily transferable to those outside that community.

The rediscovery of the importance of employees' knowledge coincided also with a popularisation of the idea of the 'knowledge worker'. This is based on the notion that certain types of work are more knowledge-intensive than others, and it is this knowledge-intensive work that is growing within the economy. Drucker coined the phrase 'knowledge workers' in 1969 (Drucker, 1969). Later Reich (1991) referred to the rise of 'symbolic analysts' – distinguishing those who deal with concepts from those who work with physical materials. In all of these post-industrial accounts, knowledge processes are argued to be intensifying, and knowledge-intensive work outgrowing traditional employment.

Conversely, given that it is inconceivable that we could have human activity without knowing and knowledge, perhaps all organisational processes involving humans are knowledge processes. It depends on what kinds of knowledge we place greatest value upon. It may be argued that all activity in organisations is 'knowledge based' to some extent, and therefore all workers are 'knowledge workers', up to a point, and all tasks performed by humans are essentially 'knowledge work'. Certainly in the 1990s many organisations rediscovered that they had underestimated employees' knowledge at a high cost, at the same time as acknowledging that they did not know what knowledge their employees had.

3.   Accelerating change in markets, competition and technology, making continuous learning essential

The gathering pace of change in most if not all sectors of the economy is a powerful source of anxiety for managers and organisations. Change occurs across several dimensions: changes in markets and industries, new forms of competition and new entrant competitors, globalisation in markets and supply chains, and changes in technology which result in product and process innovation.

Such endemic change demands continuous regeneration and development of organisational knowledge; that is, organisations and the people within them must be continually learning. The need to continually reinvent your organisation through learning is a key feature of KM. Knowledge from past experience becomes embedded within systems and technologies and embodied in routines. Continuous change requires the development of organisational routines and an organisational culture that supports the ability to create, absorb and assimilate new knowledge, and to abandon outmoded knowledge and routines. Organisations need to achieve two potentially conflicting objectives: first, to build their knowledge bases cumulatively and learn from past experience; and second, to ensure that they are learning beyond their core areas, generating the capability to assimilate new knowledge in order to be able to respond to change. These are key KM challenges.

4.   The recognition that innovation is key to competitiveness, and depends on knowledge creation and application

In many sectors competitive advantage increasingly occurs through innovation, whether in products, processes or services. The management of innovation is essentially about the management of knowledge – the creation, reformulation, sharing and bringing together of different types of knowledge. Innovation is defined in terms of the new or the novel, the break with the past, with change, and it therefore depends on knowledge creation as well as application. Knowledge is an input to innovation, inseparable from the innovation process, and new knowledge is also an output of that process.

A central KM dilemma for organisations seeking to innovate focuses on the need to build their knowledge bases cumulatively, share knowledge and learn from past experience, while at the same time to generate variety in their knowledge base as the source of novel developments. For all organisations the KM challenges presented by innovation focus on this tension between linear and non-linear processes, i.e. between the predictable and the unpredictable. Innovation therefore is far from safe and risk-free, and requires eclectic and diverse approaches to the types and sources of knowledge that might be relevant. Innovation requires the generation and incorporation of *variety* in the development of knowledge (Kogut, 2000), and making sense of knowledge across a broad spectrum.

Learning takes place throughout the innovation and diffusion processes, as problems are tackled or knowledge and expertise are shared and integrated. This new knowledge may then be incorporated in the innovation, or may generate ideas for new projects and further innovation. Cumulative and path-dependent specialised development is essentially a linear response to the challenge of innovation. Whereas continuous improvement may conform to linear development characteristics, innovation with any pretentions to novelty has by definition non-linear characteristics.

5. The growing importance of cross-boundary knowledge transactions

Arguably no firm has ever been independent in knowledge terms, but it is certainly the case today that all organisations are increasingly dependent on external sources of knowledge. The complexity and pace of change in markets and technologies make it impossible even for the largest organisations to cover all potential developments and to grow knowledge capabilities across all aspects of their operations. This is exacerbated by the unpredictable nature of change in many sectors, from manufacturing to services.

Cross-boundary knowledge transactions apply to boundaries within organisations, between functional specialisms and between disciplines. Increasingly knowledge is accessed and shared across cultural and national boundaries as organisations and markets become international. Much new knowledge – the vast majority for most firms – is created outside the corporate boundary, so organisations must develop absorptive capacity: the capability to access and assimilate new knowledge from external sources. Knowledge interdependence creates new management challenges resulting from the risks and difficulties of knowledge transactions across boundaries. So, too, the development of new products, systems and services increasingly requires the integration of knowledge from many disciplines. Alliances, networks and collaborations provide the means by which firms can reduce the risk and share costs and scarce resources, especially with regard to new or currently 'peripheral' technology areas (Quintas and Guy, 1995). However, the ability to share knowledge across functional and disciplinary boundaries presents particular KM challenges since different communities and disciplines may have little common ground for shared understandings. Such 'division of knowledge' is explored in Quintas (2002).

6. Technology limits and potentials: the limits of information technology and the potentials of communications and knowledge technologies

'We're overrun with information, but we're dying for lack of knowledge.'

(Strategic Planning Director of Qantas,
quoted in Baumard, 1999, p. 133)

It is ironic that there is a strong tendency for the information technology focus to dominate much of the current writings and conference proceedings on the subject of KM, and indeed products and service offerings marketed under the KM banner. It is ironic because, by definition, information technology is concerned with information and not knowledge *per se*. While it may sensibly be argued that codified knowledge is also information, much knowledge cannot be codified and remains inaccessible to information technology. One common delusion underpinning some KM approaches is that we can move seamlessly from data processing to information management and then to KM. As Spender (1996) points out, there is little point in introducing such a complex concept as knowledge into management discourse if we are not to take seriously the characteristics of knowledge that make it special and distinguishable from information.

There has been an increasing awareness that information systems do not capture the knowledge or even the information that managers use in their daily lives, as noted by the former head of Information Technology research for Ernst & Young:

'Evidence from research conducted since the mid-1960s shows that most managers don't rely on computer-based information to make decisions . . . managers get two-thirds of their information from face-to-face or telephone conversations; they acquire the remaining third from documents, most of which come from outside the organisation and aren't on the computer system.'

(Davenport, 1994, p. 121)

This leads to the notion that organisational information systems reached some kind of limit in the early 1990s, and need to go through a step-change in order to support KM. Certainly information technology cannot (by definition) deal with tacit knowledge, and many organisations have learned to their cost that technology should not drive any KM strategy (e.g. Thornton, 2001).

However, technology also has a number of potentials as a medium of communication, reducing time and distance constraints and, in addition, having the potential to add value to communications processes. This shifts the emphasis to using technology to support 'connectivity' rather than 'knowledge capture'. Clearly the worldwide web provides access to information on a global basis, but we need the knowledge skills of sense-making and learning if the web is to contribute to our knowledge resources and processes. Claims are also made for the KM capabilities of emerging systems, particularly in the area of communications for knowledge sharing and (in advanced systems) representation of knowledge and knowledge processes. We are at a juncture where the limitations of current information systems have been noted and the potentials for *knowledge* systems have yet to be realised.

2.2.2 The knowledge agenda

We can see that no single factor is responsible for the surge in interest in KM. What appears to have occurred is that a number of factors have come together to place knowledge high on the agenda for all types of organisations. These drivers provide a dynamic to the KM phenomenon. This provides us with a first cut at an agenda for the management of knowledge, as shown in Figure 2.2.

Of course this agenda and the drivers that underpin it will be experienced differently by different organisations, in different sectors and indeed in different global locations. We can see differing priorities in the variety of responses by firms in terms of the focus of their KM initiatives. Below we

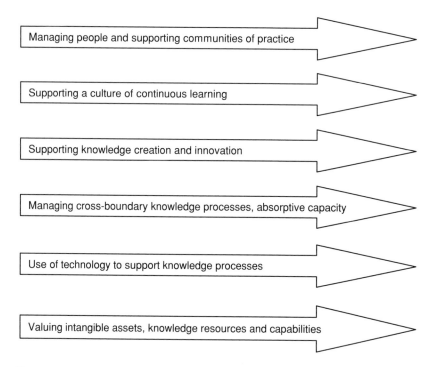

Managing people and supporting communities of practice

Supporting a culture of continuous learning

Supporting knowledge creation and innovation

Managing cross-boundary knowledge processes, absorptive capacity

Use of technology to support knowledge processes

Valuing intangible assets, knowledge resources and capabilities

Figure 2.2   Agenda for managing knowledge

take a look at some of the key processes that require attention. Before doing so we explore in a little more detail the nature of knowledge.

2.3   The nature of knowledge

The advantage of introducing the concept of knowledge into management practice is that, providing we take on board the singular characteristics of knowledge, it leads us to view organisations in a different way and to question assumptions that might otherwise be taken for granted. Of course, if we treat knowledge as an unproblematic commodity which can be packaged, traded and assimilated without difficulty then we will tend to miss the point. The disadvantage of introducing knowledge into our discourse is that the subject is complex and less than wholly amenable to simple solutions. In this short chapter we cannot hope to do more than hint at the knowledge issues that are important and relevant. We therefore focus on three key issues at the root of much of the complexity for organisations – the tacit (or implicit) nature of human knowledge, the social nature of knowledge, and the 'stickiness' or context specificity of much knowledge.

2.3.1   *Tacit knowledge*

Some forms of human knowledge can be communicated to others through language or symbols, such as the laws of thermodynamics or the names of the star constellations. Once codified, such knowledge is information or data that may be interpreted by others. Explicit or codified knowledge may be understood by people with complementary knowledge who can extract meaning from the 'codes'. Even this process of understanding or extracting meaning from information involves the use of tacit skills of interpretation, evaluation and generally making sense of what is being conveyed.

However, there is a dimension of knowledge that remains tacit and cannot be communicated in language or symbols. Polanyi (1966) uses the example of riding a bicycle. We can describe what we do when we ride, but this information would not enable another person to ride; they would have to physically ride a bicycle and learn to balance, coordinate pedalling and steering, and 'get the feel of it'. In all human activity we acquire tacit knowledge through experience and internal reflection, and this is impossible to communicate or share with others who have never been through similar learning experiences. This is at the heart of some of the major challenges for KM in organisations.

The key challenge can be illustrated by analogy. Imagine a group of jazz musicians who come together for the first time. All are accomplished musicians, each a virtuoso on their own instrument, but specialists, not multi-instrumentalists. They know how to play their instrument because they have practised and internalised experiential knowledge. All are experienced improvisers who have performed with many different line-ups. They share a knowledge of the jazz canon, being steeped in listening to the great jazz musicians over many years. They know how to play in a group and react to their fellow musicians' expressions, exchanging phrases and changing the tempo in an interwoven pattern that is created spontaneously.

There are many learning opportunities in such a scenario, and indeed each musician may well take away with him or her new insights, phrasings, interpretations and other learning points about group interaction. However, even with this level of empathy and shared reference points, the piano player cannot go home and play the saxophone like the virtuoso colleague in the group, and vice versa. Neither could they learn each other's instrument by having extended conversations over the next six months about technique and method. In order to acquire knowledge of a new instrument, they would each have to practise. Such experiential knowledge cannot be codified – it remains tacit. In a sense, what they possess is the ability to enact *knowing*, rather than knowledge; it is a set of cerebral and motor-neuron internalisations, which only have reality when they play. Should anything happen to their hands, where then is their knowledge?

Similarly, organisations survive, innovate and prosper because people within them generate knowledge, and a proportion of this is experiential and must remain tacit. This awareness points to the limitations of attempts to manage knowledge based on that which is codified or codifiable. It is a simple point, and has profound implications, but it is one that is all too easily glossed over.

2.3.2 Social knowledge and communities of practice

The jazz musician analogy illustrates the tacit and personal nature of experiential, tacit knowledge. However, it also showed how the musicians could learn from each other, up to a point. As has been shown by Spender, Brown, Wenger, Baumard and others, knowledge has a social dimension – it may be created and held collectively. People who share work experiences, problem agendas and have similar learning opportunities may be said to form communities of practice (CoPs) (Lave and Wenger, 1991). Wenger (2000) defines a CoP as a social learning system, united by joint enterprise, mutually recognised norms and competence, with shared language, routines and stories. The example of CoPs illustrates the long history of the management of knowledge without the KM label.

A CoP is most often an informal grouping. It may be unrecognised (Scarbrough, 1996) or ignored or taken for granted (Baumard, 1999) in the organisation. So too, it may transcend organisational boundaries, including people in several organisations who hold experiences in common. CoP members act as resources for each other, 'exchanging information, making sense of situations, sharing new tricks and ideas' (Wenger, 1998, p. 47). In Xerox, photocopier engineers were observed working together on a problem machine, communicating like jazz musicians, exchanging truncated phrases and able to communicate non-verbally because of shared experience, shared learning, shared understandings (Brown and Duguid, 1991). CoPs therefore represent oases within which knowledge processes function naturally.

The challenge posed by social knowledge is that it may not be acknowledged by management. Formal management styles may be at odds with the informality of CoPs and indeed attempts to formally manage CoPs from outside may undermine them. Baumard (1999) identifies three CoPs in the Australian airline Qantas: the pilots and their retinue, the financial group and the marketing group. Each of these communities has their own language, which, as Baumard emphasises, indicates different interpretations of reality. Qantas' top-down management style favours documents, manuals and computerised information, whereas the CoPs favour less explicit circulation of knowledge: '...communities of practice, conjectural knowledge and repertories of thought inscribed in practice are all tacit' (Baumard, 1999, p. 135). The Qantas communities refused to use a new computer-based 'knowledge management system' introduced from outside the CoPs.

2.3.3   Context specificity and 'stickiness' of knowledge

As we have seen, much of the knowledge generated in organisational processes is tacit knowledge that is held implicitly by those directly involved. Skilled members of a community of practitioners are often unaware of the tacit knowledge they possess, e.g. their problem recognition and problem solving behaviour, the rules that they follow and the knowledge sources that they draw on. Moreover, communication among community members is often dependent on implicitly held knowledge of community-shared interpretations, beliefs and culture.

Tacit knowledge is personal, linked to experience and learning, and cannot be codified. In some situations tacit knowledge may, however, be shared (via non-codified pathways) within groups that share common learning experiences and understandings rooted in common practice (Nonaka and Takeuchi, 1995; Brown and Duguid, 1998). It may also be embedded in the organisational and social processes that provide the context in which it is generated.

Current perspectives on knowledge suggest that knowledge is created in specific contexts and is to varying degrees 'situated' (Lave, 1993) and may be 'sticky' (von Hippel, 1994) and difficult to transfer or share. Much of the knowledge generated through, for example, research and development (R&D) is of a tacit nature and located in the specific context in which it was developed (Nelson and Winter, 1982). Tacit organisational processes are difficult to transfer or share between contexts, as is illustrated by the example of Chaparral Steel. The CEO is happy to tour competitors through the Chaparral plant, showing them 'almost everything and we will be giving away nothing because they can't take it home with them' (Leonard, 1995, p. 7).

What has value and meaning in one context may have little or no meaning in another context. It follows that in order to transfer knowledge between contexts it must be decontextualised, or made independent of context. Indeed, many authors argue that the only way to transfer tacit knowledge is through guided joint social interaction during, for example, apprenticeship within a community of expert practitioners (Lave and Wenger, 1991). Similarly, where knowledge associated with technology is being transferred between organisations, 'in the absence of intimate human contact, technology transfer may sometimes be impossible' (Teece, 1981, p. 86).

There are significant cultural differences in the extent to which knowledge may be shared between contexts. In a revealing study of collaborative projects involving Western and Japanese firms, Hamel *et al.* (1989) found that Western companies tended to bring easily imitated technology to a collaboration, whereas Japanese firms' strengths were often 'difficult to unravel' competencies which were less transferable. This leads us to move across organisational boundaries and consider the extra-organisational dimensions of KM.

2.4   Extra-organisational knowledge and absorptive capacity

In terms of knowledge resources and capabilities, few organisations would consider themselves to be self-sufficient islands. However large the firm, you can be fairly sure that more potentially relevant knowledge is generated outside the boundaries than inside. Knowledge may be created in universities and research institutes, in customers, suppliers or competitors in the firm's own sector, or indeed in other sectors. Such awareness underpins current emphasis on the importance of *absorptive capacity* or the organisations' ability to *learn* from external sources. As Cohen and Levinthal (1990) argue, absorptive capacity requires the investment of resources. Indeed, one of the primary reasons why firms invest significant resources in research and development is in order to track and understand external developments (Cohen and Levinthal, 1989).

The question that arises is how far should firms develop, as part of their capacity to absorb knowledge, capabilities in the areas of knowledge they wish to acquire? When does absorptive capacity become an expanded internal knowledge base? Loasby (1999) argues the case for the superior advancement of knowledge through specialisation, and is clear that firms should not attempt to widen their knowledge capabilities too broadly:

'Many of the capabilities which a firm requires must be left outside its control if they are to continue to develop. Capabilities must remain distributed; what the firm needs is access to these capabilities.... all firms, and all individuals, require absorptive capacity if they are to make good use of capabilities and information which is generated elsewhere.'

(Loasby, 1999, p. 96)

To do this firms must be open to other ways of thinking; to be agile and receptive. The skills of absorption include techniques of sourcing, sense-making and acquisition. They must become adept at learning. However, much depends on the sector, the stage of the lifecycle of the underpinning technology and product markets, and the structural nature of relationships with external sources.

A firm's capacity to track and to absorb external knowledge, however, is constrained by what it already knows (Pavitt, 1991, 1998; Patel and Pavitt, 1997). The knowledge base of a firm is developed cumulatively and is path-dependent, constraining what can be learned. Even large firms that have capabilities across a number of fields constrain their search activities close to what they already know: 'In this sense, there are clear cognitive limits on what firms can and cannot do' (Pavitt, 1998, p. 441).

This path dependency may be an appropriate incubator for continuous improvement and incremental innovation, that is in stable markets, but less so in volatile markets where there are step changes and more radical innovation. Where markets and products are relatively stable and trajectories of development predictable, these cognitive constraints may present less of a risk. Where, as is the case in many sectors and markets, change is less predictable, firms need to widen their knowledge base in order to be able to innovate. Cognitive constraints imposed by core capabilities may indeed become core rigidities (Leonard, 1995).

2.5 Key knowledge processes

We can now summarise the array of key processes that emerge from the above. These are the processes that organisations must address if they have an aspiration to manage knowledge more effectively.

We previously identified the key drivers of KM which scoped the main dimensions of the phenomenon (see Figure 2.2). Six dimensions were discussed, leading to the beginnings of an agenda for current practice:

(1)   valuing intangible assets, knowledge resources and capabilities
(2)   managing people and supporting communities of practice
(3)   supporting a culture of continuous learning

(4) supporting knowledge creation and innovation
(5) managing cross-boundary knowledge processes, absorptive capacity
(6) use of technology to support knowledge processes.

This begins to scope the field, but such high-level descriptors are under-pinned by a number of processes that cut across these themes. Underlying all of the above are the generic processes of communication, learning and thought, without which human activity cannot function. We can also identify another layer of organisational processes:

● creating/generating/producing knowledge
● sharing knowledge
● sourcing knowledge
● sense-making
● synthesising/transforming/combining knowledge
● capturing/storing/classifying knowledge (as information)
● mapping knowledge or knowledge proxies
● measuring knowledge or knowledge proxies
● applying and reusing knowledge
● managing knowledge processes across boundaries.

The use of 'proxies' in the above indicates that knowledge itself may not be amenable to mapping or measurement, but proxies may be used to provide some indication of the presence of knowledge. For example, it may not be possible to map or measure the knowledge within a community of practice other than by its outputs. So too, patents and copyright may provide some measure of the capability for knowledge creation.

Finally, we can see that key capabilities discussed above such as absorptive capacity require a combination of sourcing, sense-making, learning, synthesising and combining knowledge.

How do organisations turn the above into practice? There are differing priorities in the variety of ways that firms approach their KM initiatives. For the majority of firms in the West, the priorities are the 'capture' of employees' knowledge, exploitation of existing knowledge resources or assets, improved access to expertise (i.e. improved 'know-who'), transferring knowledge between projects, and building and mining knowledge stores. Like many organisations, UK Post Office Consulting launched a cluster of KM projects in the late 1990s, including initiatives focused on:

● knowledge sharing (targeted on communicating, learning, reviewing, capturing and sharing knowledge)
● use of stories to communicate experience (targeted on transferring learning)
● after-action reviews (capturing learning from experience)
● use of intelligent agents (identifying specific and tailored information or contacts)

- developing a people database (providing access to expertise)
- using expert interviews (to capture expertise)
- learning from mistakes (surfacing and capturing learning in a non-blame culture, avoiding costly repetition)
- using expert masterclasses (to share expertise).

(Quintas *et al.*, 1999)

2.6 Conclusions

Significant developments in the economy and business environment at the end of the 20th century prompted organisations of all types to rethink the nature of the resources and capabilities that generate advantage. For the first time the concept of KM came to the top of the management agenda. This is perhaps more remarkable for the previous neglect rather than the current interest. However, it does not take long to realise that knowledge processes have been managed for as long as organisations have existed, although without the KM label.

The introduction of *knowledge* into our thinking about organisations poses additional and qualitatively different questions from that of an informa-tion agenda. Unfortunately knowledge cannot be regarded as a commodity that is easy to manage, trade and share. Knowledge is created and applied through dynamic processes. *Knowing* is an active process, resulting from action and engagement (Cook and Brown, 1999). 'Knowledge can no longer be pinned down to the heads of individuals and treated as a finished, sta-ble product but is instead to be seen as a relational, transient product...' (Araujo, 1998, p. 324, building on Lave, 1993 and Lave and Wenger, 1991). Much knowledge is experiential and known to the individuals involved in specific activities. Experiential knowledge may be difficult if not impossible to communicate to others – much of it remains tacit.

Knowledge is created in specific contexts and to an extent is dependent on *context* to acquire meaning (Lave and Wenger, 1991; Lave, 1993). Context-specific (or *situated*) knowledge may often be transient. Certain types of knowledge are 'sticky' and very difficult to share between contexts, even within a single organisation.

Knowledge has a social dimension. Communities of practice within which individuals share common work experiences and problem agendas pro-vide a social context within which knowledge may be created and effec-tively shared (Brown and Duguid, 1998). Within the community of practice, tacit knowledge may be shared in non-codified forms (Brown and Duguid, 1998). Such communities have a degree of exclusivity – their knowledge is not readily available to outsiders who do not share similar practice-based experiences.

Again and again we find that organisational culture determines knowl-edge practices. In this regard many Western organisations in particular may begin from disadvantaged positions, because of the tendency to privilege codified knowledge or information. However we approach it, the challenges are significant. The processes of creating, acquiring and applying knowledge require learning and understanding, which presupposes capability – the investment of time and resources. Perhaps the one thing it has in common with all other commodities is that knowledge is not a free good.

This chapter has summarised the scope of KM and identified a range of processes that constitute the beginnings of a management agenda for action. Our focus has intentionally been broad. The following chapters develop aspects of these themes in the specific context of the construction industry.

References

Araujo, L. (1998) Knowing and learning as networking. *Management Learning*, **29**(3), 317–36.

Baumard, P. (1999) *Tacit Knowledge in Organisations*. Sage, London.

Bell, D. (1973) *The Coming of Post-Industrial Society*. Heinemann, London.

Brown, J.S. and Duguid, P. (1991) Organisational learning and communities of practice: towards a unified view of working, learning and innovation. *Organisation Science*, **2**(1), 40–57.

Brown, J.S. and Duguid, P. (1998) Organizing knowledge. *California Management Review*, **40**(3), 90–114.

Brown, J.S. and Duguid, P. (2000) *The Social Life of Information*. Harvard Business School Press, Boston, Massachusetts

Cohen, W.M. and Levinthal, D.A. (1989) Innovation and learning: two faces of R&D. *Economic Journal*, **99**, 569–96.

Cohen, W.M. and Levinthal, D.A. (1990) Absorptive capacity: a new perspective on learning and innovation. *Administrative Science Quarterly*, March, 128–52.

Cook, S.D.N. and Brown, J.S. (1999) Bridging epistemologies: the generative dance between organisational knowledge and organisational knowing. *Organisation Science*, **10**(4), 381–400.

Cross, R. (1998) Managing for knowledge: managing for growth. *Knowledge Management*, **1**(3), 9–13.

Davenport, T.H. (1994) Saving IT's soul: human-centred information management. *Harvard Business Review*, March–April, 119–31.

Drucker, P.F. (1969) *The Age of Discontinuity: Guidelines to Our Changing Society*. Heinemann, London.

Hamel, G., Doz, Y.L. and Prahalad, C.K. (1989) Collaborate with your competitors – and win. *Harvard Business Review*, Jan–Feb, 133–9.

Hammer, M. (1990) Re-engineering work: don't automate, obliterate. *Harvard Business Review*, July–Aug, 104–12.

Hippel, E. von (1994) 'Sticky information' and the locus of problem solving: implications for innovation. *Management Science*, **40**(4), 429–39.

Kogut, B. (2000) The network as knowledge: generative rules and the emergence of structure. *Strategic Management Journal*, **21**(3), 405–25.

Lave, J. (1993) The practice of learning. In *Understanding Practice. Perspectives on Activity and Context* (S. Chaiklin and J. Lave, eds). Cambridge University Press, Cambridge.

Lave, J. and Wenger, E. (1991) *Situated Learning: Legitimate Peripheral Participation.* Cambridge University Press, Cambridge.

Leonard, D. (1995) *Wellsprings of Knowledge: Building and Sustaining the Sources of Innovation.* Harvard Business School Press, Boston.

Loasby, B.J. (1999) *Knowledge, Institutions and Evolution in Economics.* Routledge, London.

Machlup, F. (1962) *The Production and Distribution of Knowledge in the United States.* Princeton University Press, Princeton New Jersey.

Marshall, A. (1972) *Principles of Economics*, 8th edition, first published 1890. Macmillan, London.

Massey, D., Quintas, P. and Wield, D. (1992) *High Tech Fantasies: Science Parks in Society, Science and Space.* Routledge, London.

Murray, P. and Myers, A. (1997) The facts about knowledge. *Information Strategy*, Sept, 31–3.

Nelson, R.R. and Winter, S.G. (1982) *An Evolutionary Theory of Economic Change.* Belknap Press, Cambridge, Massachusetts.

Nonaka, I. and Takeuchi, H. (1995) *The Knowledge-Creating Company: How Japanese Companies Create the Dynamics of Innovation.* Oxford University Press, Oxford.

O'Dell, C. and Grayson, C.J. (1998) If only we knew what we know: identification and transfer on internal best practice. *California Management Review*, **40**(3), 154–74.

Patel, P. and Pavitt, K. (1997) The technological competencies of the world's largest firms: complex and path-dependent, but not much variety. *Research Policy*, **26**, 141–56.

Pavitt, K. (1991) Key characteristics of the large innovating firm. *British Journal of Management*, **2**, 41–50.

Pavitt, K. (1998) Technologies, products and organisation in the innovating firm: what Adam Smith tells us and Joseph Schumpeter doesn't. *Industrial and Corporate Change*, **7**(3), 433–52.

Penrose, E.T. (1959) *The Theory of the Growth of the Firm.* Basil Blackwell, Oxford.

Polanyi, M. (1966) *The Tacit Dimension.* Routledge & Kegan Paul, London.

Ponzi, L. and Koenig, M. (2002) Knowledge management: another management fad? *Information Research*, **8**(145) (available at http://informationr.net/ir/8-1/paper145.html; accessed 21 Sept 2004).

Quintas, P. (2002) Implications of the division of knowledge for innovation in networks. In *Networks, Alliances and Partnerships in the Innovation Process* (J. de la Mothe and A.N. Link eds). Kluwer Academic Press, Boston, pp. 135–62.

Quintas, P. and Guy, K. (1995) Collaborative, pre-competitive R&D and the firm. *Research Policy*, **24**, 325–48.

Quintas, P., Jones, J. and Demaid, A. (1999) *An Introduction to Managing Knowledge.* The Open University, Milton Keynes.

Reich, R. (1991) *The Work of Nations: Preparing Ourselves for 21st-Century Capitalism.* Simon and Schuster, London.

Roos, J. and Roos, G. (1997) Valuing Intellectual Capital. *FT Mastering Management*, 3, July–Aug, 6–10.

Scarbrough, H. (1996) *Business Process Re-design: The Knowledge Dimension.* ESRC Business Processes Resource Centre, University of Warwick.

Skandia (1994) *Visualizing Intellectual Capital in Skandia.* Supplement to Annual Report, Skandia, Stockholm, Sweden.

Spender, J.C. (1996) Making knowledge the basis of a dynamic theory of the firm. *Strategic Management Journal*, **17**, Winter Special Issue, 45–62.

Sveiby, K.E. (1997) The intangible assets monitor. *Journal of Human Resource Costing and Accounting*, **2**(1), 25–36.

Systematic (2002) Intellectual Capital Report. http://www.systematic.dk/uk/about+us (accessed 21/07/2004).

Teece, D.J. (1981) The market for know-how and the efficient international transfer of technology. *Annals of the American Academy of Political and Social Science*, **458**, 81–96.

Thornton, S. (2001) Knowledge management in Dstl. Presentation at *Knowledge Management in Practice* Workshop, The Open University, Milton Keynes, 12 October.

Wenger, E. (1998) *Communities of Practice.* Cambridge University Press, Cambridge.

Wenger, E. (2000) Communities of practice and social learning systems. *Organisation*, **7**(2), 225–46.

3 Construction as a Knowledge-Based Industry

Charles O. Egbu and Herbert S. Robinson

3.1 Introduction

The emergence of the knowledge era as an integral part of the global economy is leading to dramatic changes in the business environment. Knowledge management (KM) and its manifestation in the expertise of people is now seen as the greatest value of creation for organisations. In a recent Competitiveness White Paper, *Our Competitive Future: Building the Knowledge Driven Economy*, a knowledge economy was defined as one in which the generation and the exploitation of knowledge have come to play a predominant part in the creation of wealth (DTI, 1998). Similarly, the OECD report *The Knowledge-Based Economy* suggested that what is created in a knowledge economy is a network society, where the opportunity and capability to access and join knowledge- and learning-intensive relations determine the socio-economic position of individuals and firms (OECD, 1996). The issues of knowledge production, transmission and transfer are important facets of the knowledge economy.

In a knowledge economy, it could be argued that different kinds of knowledge are evident. These include:

- *'Know-what'* – accumulation of facts that can be broken down into pieces.
- *'Know-why'* – scientific knowledge that underlies technological development, product and process advancements.
- *'Know-how'* – skills or capability to do something and the reason for the formation of industrial networks to enable firms to share and combine elements of know-how.
- *'Know-who'* – involves information about who knows what and who knows how to do what.

Other characteristics of a knowledge economy include an intensified knowledge codification, accelerated transmission of information and emergence of flexible organisations. Such organisations are characterised by multi-task responsibilities, teamwork, job rotation to achieve high product quality and some customisation, together with the speed and low unit cost of mass production. The implications are:

- An increased demand for knowledge, skills and learning; the formation of hierarchies of networks driven by the acceleration of the rate of change and rate of learning.
- Growth of learning organisations that search for linkages to promote interfirm interactive learning, and for partners and networks to provide complementary assets, forming innovation systems and business clusters; formation of complex chains of creation, production and distribution, and global competition.

This chapter discusses knowledge production, transmission and transfer in the context of the construction industry. Following this Introduction, Section 3.2 outlines key characteristics of the construction industry and its role in delivering knowledge-intensive products and services. It is argued that the construction industry, although known for its highly tangible products such as buildings and other structures, is increasingly now recognised as a provider of services, placing more emphasis on knowledge. Section 3.3 focuses on the knowledge production process, the interaction of tacit and explicit knowledge in the creation of new knowledge, and the type of knowledge to manage in construction processes, people and products. Issues relating to skills and competencies in supporting knowledge production are also discussed. Section 3.4 deals with mechanisms facilitating the transmission and transfer of knowledge. The importance of communicating and sharing knowledge within and across organisational boundaries to cope with the increase in collaborative working practices is highlighted. Section 3.5 discusses the need for creating and sustaining a knowledge culture to facilitate knowledge production, transmission and transfer. Section 3.6 presents a summary of key conclusions and recommendations.

3.2   The construction industry and knowledge-intensive products and services

Today's UK construction industry increasingly shares many of the characteristics of the knowledge economy. The industry employs in excess of 1.5 million people and contributes around 8% of GDP. It has the third largest construction output in Europe and is the fifth largest in the world. Exports are in the order of £10 billion, whilst domestically the construction industry is a major sector for delivery of key government programmes such as housing, hospitals and infrastructure. The industry is heterogeneous and diverse, consisting of different organisations, consultants, building materials and product producers, and professionals providing a range of services for clients, customers and the wider community. The industry is dominated by small and medium enterprises (SMEs), which make up over 90% of all organisations, with a relatively small number of large companies. UK consultants and contractors operate in almost every country throughout the world. The knowledge group of occupations is not broadly homogeneous. Some categories of occupations participate more closely than others in 'scientific and technological' activities, i.e. they are more knowledge-intensive.

Too often, the industry is known for its products (e.g. buildings, roads, bridges, dams and monuments) and not seen as an industry that provides services to its clients and customers. This is despite the very high levels of 'service-input' needed in the formation of construction products. Knowledge-intensive organisations rely on professional knowledge or expertise relating to a specific technical or functional domain. The term knowledge-intensive, in a way, intimates economists' labelling of firms as capital-intensive or labour-intensive. To some extent, these 'labels' describe the relative importance of capital and labour as production inputs. In a capital-intensive firm, capital has more importance than labour; in a labour-intensive firm, labour has the greater importance. By analogy, labelling a firm as knowledge-intensive implies that knowledge has more importance than other inputs.

The knowledge economy is often considered to include 'research and development'-intensive industries and advanced services, whilst too often the so-called resource-based sectors are regarded as medium- or low-knowledge activities. This position is changing fast. Whilst the resource-based organisations do not undertake extensive research and development, they now have to adopt increasing amounts of advanced technology in order to maintain a competitive advantage in an increasingly global marketplace. This is increasingly achieved by focusing on new-added products and services, which has placed an increasing emphasis on skills and knowledge.

Research conducted by Windrum *et al.* (1997) and den Hertog and Bilderbeek (1998) identified design, architecture, surveying and other construction services as knowledge-intensive service sectors. An important feature that distinguishes knowledge-intensive sectors from manufacturing firms is the type of 'product' they supply and, following this, the role they play in regional and national innovation systems. Whereas manufactured products and processes contain a high degree of codified knowledge (they are a 'comodification' of knowledge), knowledge-intensive sectors are characterised by a high degree of tacit ('intangible') knowledge. Specialised expert knowledge and problem-solving know-how are the real products of knowledge-intensive services. Construction activities can be highly knowledge-intensive. Just as 70% of the production cost of a new car can be attributed to knowledge-based elements such as styling, design and software (Scottish Enterprise, 1998), the same can be said of the building of a new modern office complex. A new modern office complex has a high proportion of its development cost attributable to knowledge-based elements such as design, an assessment of cost alternatives of different components of the building, advice on contractual aspects, risks and buildability of the project, quality, health and safety issues on the project, to mention but a few.

Professional knowledge, i.e. knowledge produced by consultants while interacting with their clients' settings, is deeply embedded in a mutual socialisation process, where consultants and their clients design together their final output. This is often seen in the kinds of services provided by professional/consultancy firms of architects, quantity surveyors and engineers. For consultancy or professional firms, their main capital is intellectual assets, and most of their processes are geared towards the exploration, accumulation and exploitation of individual and firm expertise.

3.3   Knowledge production in construction

Organisational knowledge is a mixture of tacit and explicit knowledge. Tacit knowledge is stored in individuals' heads. It is a product of experiences, insights and intuition which could be technical (i.e. know-how of an expert) or cognitive (i.e. based on values, beliefs and perceptions). In the context of construction, examples of tacit knowledge include estimating and tendering skills acquired over time through hands-on experience of preparing bids, understanding the construction process, interaction with clients/customers and project team members in the construction supply chain, as well as understanding tender markets. This type of knowledge is experiential, judgmental, context-specific and therefore difficult to codify and share.

Explicit knowledge is stored as written documents or procedures. As this type of knowledge is codifiable, it is reusable in a consistent manner and therefore easier to share. Examples of explicit knowledge in construction include design codes of practice, performance specifications, drawings in paper-based or electronic format and construction techniques. Materials testing procedures, design sketches and images, 3-D models and textbooks are also examples of explicit knowledge.

3.3.1   *Knowledge creation*

According to Nonaka and Takeuchi's (1995) theory of knowledge creation, there are four distinct modes of interaction that result in the creation of knowledge (see the SECI model in Figure 3.1). Construction project knowledge is created through the actions of individuals, project teams and

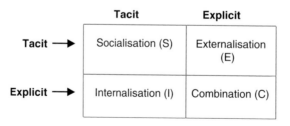

	Tacit	Explicit
Tacit →	Socialisation (S)	Externalisation (E)
Explicit →	Internalisation (I)	Combination (C)

Figure 3.1   Knowledge creation theory (Nonaka and Takeuchi, 1995)

construction organisations, and the interactions of these different types of knowledge (explicit and tacit) from concept design to handing over of the completed project.

Tacit to tacit interaction takes place through the process of socialisation. An architect giving a verbal account or an explanation of a design concept to a client during a meeting is an example of this type of interaction. Apprentice carpenters, bricklayers, plumbers, etc. often work with their masters to learn craftsmanship not through formal instruction but by socialisation which involves observation, imitation and practice. The long tradition of apprenticeship schemes in the construction industry is responsible for producing various craftsmen who rely on their tacit knowledge to solve construction problems. Such experiential knowledge is reinforced and developed through shared experience by continuous interaction and learning from each other. Similarly, young engineers, architects and surveyors supplement their academic training through mentoring. The mentors are often senior management staff who can help individuals to learn, unlock their talents and develop their knowledge in the organisation.

Internalisation takes place when knowledge is transformed from explicit to tacit by individuals. For example, an architect reading a textbook on design theory, or using a manual on design standards, could interpret these explicit documents to create an internal mental model of a unique design satisfying the client's requirements and his/her taste and style. Externalisation is the reverse process where tacit knowledge is made explicit so that it can be shared. An architect engaged in a discussion with a contractor on site, which is subsequently followed by a written instruction made available to specialist subcontractors, engineers and quantity surveyors, is an example of an externalisation process. This process also takes place when an architect translates a design concept or mental model into sketches to explain to a client.

Explicit to explicit knowledge interaction takes place through a process called combination. Combination involves gathering, integrating, transferring, diffusing and editing knowledge (Nonaka and Toyoma, 2003). Individuals and project teams in construction create knowledge through integrating and processing various project documents (e.g. design brief, sketches, project programme, engineering and production drawings, performance specifications, conditions of contract, bills of quantities). Technologies such as e-mails, databases, CAD systems, document management systems and project extranets facilitate this mode of knowledge conversion.

Various other technologies and techniques are used to facilitate other knowledge conversion processes, such as face-to-face interactions, communities of practice, project review meetings, brainstorming sessions and 'toolbox talks' on site. Some of these are discussed in Chapter 6.

Much of the training and experience of construction professionals is based on a balance between codified (explicit) knowledge and tacit knowledge. Case study interviews with structural design firms show that about 80% of knowledge used during concept design is tacit compared to about 20% of explicit knowledge, whilst the reverse is true at the detailed design stage – 20% tacit and 80% explicit (Al-Ghassani, 2003). It is the dynamic interactions between tacit and explicit knowledge that facilitates decision making in the implementation of construction projects. This is why construction project documents are understood and interpreted by those who have been through the same or similar type of training. For example, structural engineers can extract meanings from design codes or interpret construction drawings easily whilst accountants cannot. It was highlighted in the previous chapter that managers get two-thirds of their information from face-to-face or telephone conversations (tacit) and the remaining third from documents (explicit).

3.3.2   Implications for knowledge management

Decisions on what knowledge a construction organisation needs or the knowledge intensity depends on the context of the business environment, i.e. key knowledge about processes and people for the delivery of its products. These context-based factors address issues of what is produced (products – goods/services), how it is produced (processes) and by whom (people).

There are therefore three aspects of knowledge to manage in the construction context: (1) products or project types, (2) processes and (3) people (see Figure 3.2). The knowledge base of construction organisations is a function of the procedures put in place to capture knowledge about processes, products, as well as people, because knowledge primarily resides in people, not technology (Davenport, 2000). Technology supports connectivity; it is therefore an important enabler to support the KM process.

Product-based factors relate to the characteristics of the services or goods to be produced, whether standardised, mature or innovative (Hansen *et al.*, 1999). Construction project organisations are characterised by the types

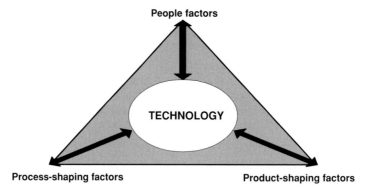

Figure 3.2   Context-based factors influencing a KM strategy. (Reproduced with permission from Robinson *et al.*, 2001)

of projects or the products they deliver. There is a range of 'component products' to produce different types of 'end products', from small and simple buildings to large and sophisticated structures such as the Millennium Bridge. Construction 'end products' are classified into three distinct types: standard construction, traditional construction and innovative construction (Bennett, 1991). Innovative projects are needed to satisfy the demands of clients with unusual needs, or where established answers are no longer appropriate as a result of market or technological changes (Bennett, 2000). Knowledge about clients, end-users and market characteristics is therefore important for construction organisations. Clients range from 'one-off' or occasional clients with very little knowledge of construction to major clients with repeat business and considerable knowledge of construction. End-users may have varying needs and aspirations too. The type of knowledge to be managed in construction is therefore influenced by a combination of client, end-user and market characteristics.

Process-based factors relate to the technical and management systems required for the delivery of products. The 'end products' required by clients are often different as are the (technical and management) processes used in the delivery of construction projects. This difference has profound implications for processes and the types of knowledge to be managed during design and construction. Technical processes range from highly knowledge-intensive approaches relying mainly on tacit knowledge, such as producing design sketches using pencil and paper, to automated processes relying on intelligent and knowledge-based systems (explicit knowledge) codified in plant, machinery or robots for on-site construction. The management processes depend on the type of product to be delivered. Standard construction products are more effectively managed by programmed organisations relying heavily on routine and standard procedures (codified knowledge) to manage the design and construction process. Traditional construction products require professional organisations relying on a mix of standard and flexible procedures (tacit and explicit knowledge) to manage the design and construction process, with specialist contractors who are able to interpret design information in order to manufacture and construct the products. Innovative construction requires highly flexible management procedures characterised by a higher utilisation of tacit knowledge to manage complex design and construction processes. New knowledge is developed through problem solving or creativity to find answers to fulfil unusual and demanding client requirements.

People-based factors relate to skills, problem-solving abilities and the characteristics of teams. Bennett (1991) argued that while appropriate project management structures are necessary to tackle the different types of products, they are not sufficient to ensure an efficient construction industry. Highly skilled individuals and competent teams (designers, suppliers and constructors) are vital for the construction process. Standard construction requires individuals with basic knowledge and skills.

However, problem-solving or creative people are needed for innovative projects that are often ill-defined and complex to implement. Individuals with tacit knowledge are central to the creativity required in the design and construction of innovative projects. Team stability also has a profound implication for knowledge creation and reuse. Egan (1998) noted that 'the repeated selection of new teams inhibits learning, innovation and the development of skilled and experienced teams'. This view is supported by Bennett (2000) who argued that the best result comes from the same people working together project after project.

3.3.3 Knowledge mapping

The starting point for structuring construction project knowledge is to develop a knowledge map for locating explicit knowledge, and for serving as pointers to holders of tacit knowledge. Figure 3.3 shows a knowledge map with multiple level of details. The items or elements on the knowledge map can be text, drawings, graphics, documents, directories, icons, symbols or models which can also serve as links to more detailed knowledge. For example, skills yellow pages are widely used to provide a directory of experts. This can help in finding the right person to approach for advice and best practice. A leading consulting organisation has skills yellow pages that put one in contact with not just another person but that individual's network (Sheehan, 2000). Such a knowledge mapping tool is very important but needs to be kept up to date to maintain its usefulness.

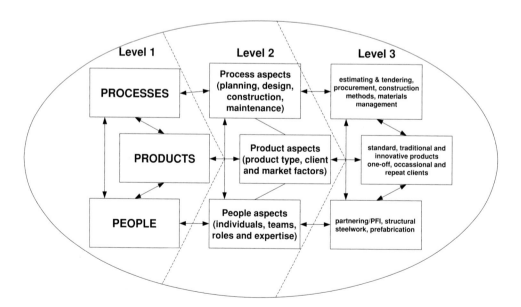

Figure 3.3   Knowledge map (Robinson et al., 2002) (PFI, private finance initiative)

Table 3.1 Examples of process knowledge base

Sub-processes	Key knowledge issues
Procurement	Partnering, PFI, design and build, construction management, traditional contracting
Estimating and tendering	Profit margins, overheads, bidding success rate, bidding costs, regional factors, sub-contract quotations
Materials management	Structural steelwork, concrete
Construction methods	Prefabrication, on-site construction

Source: Robinson *et al.* (2002)

Table 3.1 illustrates a range of process knowledge areas that construction organisations may wish to explore. A similar profile could be developed for people and product knowledge.

The knowledge map serves as a continuously evolving project memory, forming a link between different knowledge sources, capturing and integrating new knowledge into the project knowledge base. It also enables construction project team members to learn from past and current projects through the navigation of information as well as the creation of new knowledge, by adding, refining and broadening the scope.

3.3.4 Skills and competencies in a knowledge economy

In order to truly embrace the ethos of the knowledge economy, the construction industry and its firms will need to address a host of challenges. The main challenges include having to cope with increasing competition, the construction market becoming more global, the changing levels and patterns of demands from clients, customers and the society, and dealing with the pace and implications of changes in information and communication technologies (ICT). These have important ramifications for organisations, some of which are structural as well as cultural. For many organisations, the knowledge economy involves rapid change, uncertainty and turbulence. Firms will need to continuously adapt their technical and management processes, seek and retain appropriate skills and competencies to deal with unusual product demands, and accommodate new technologies and grasp new opportunities.

In order to have a competitive industry, and one that is capable of improvement in a knowledge economy, it is essential that the industry has an efficient, motivated and competent workforce. In such an economy, employee mobility and loyalty are important issues for management. A future where 50% of school leavers are expected to proceed to higher education is one that is placing pressure on the industry's ability to recruit at operative and technical levels, and requires adjustments to recruitment at higher technical, managerial and professional levels. Competing in the labour market of the future, construction will need to be seen as a cutting-edge knowledge-based industry – providing career opportunities, better working practices and innovation. In recent years, lack of requisite knowledge and of skilled labour within the industry has come to the forefront, with clear indicators that skills are failing to keep up with demand. *The State of the Construction Industry Report* (DTI, 2002) highlights that over the last few years, intake onto construction-related degree courses has declined considerably, affecting the capacity of the industry to deliver the government's programme of infrastructure renewal and improved value for money. Future skills requirement and any skills restructuring need to be devised to facilitate the meeting of wider social and political targets demanded by clients and government, and sought by the industry. These should include policies related to working conditions, workforce engagement (including diversity and equality), health and safety, career development and life-long learning.

The key skills requiring further training were identified in the Construction Research and Innovation Strategy Panel's recent reports, *Changing Skills Needs in the Construction Industry* (CRISP, 2001) and *Culture and People in Construction – A Research Strategy* (CRISP, 2002). These skills are inter/intra-team trust, interorganisational teamwork and co-operative working. The skills associated with exploiting information communication technologies for improved business needs are also vital. In construction, there is a slow but increasing take-up of e-business approaches and models by a few practices, suppliers and manufacturers.

It is estimated in the *Accelerating Change* report (Strategic Forum for Construction, 2002) that the construction industry will need to recruit the equivalent of 76 000 workers per annum based on modest growth estimates. Given changing clients' demands and the desire to apply new technology and processes, it is essential that the key culture and people issues in construction are addressed urgently. Engagement of the workforce is vital to establishing a culture of 'organisational learning' as this enables the workforce to continually learn and update its skills. This is needed for meeting the challenges of the knowledge economy and for improved competitiveness. In addition, both the reports *A Commitment to People – Our Biggest Asset* (Rethinking Construction, 2000) and *Respect for People – A Framework for Action* (Rethinking Construction, 2002) noted that construction organisations that fail to improve their attitude and performance towards respecting people will fail to recruit and retain the best talent and business partners.

3.4   Communicating and sharing knowledge

In the construction industry, as in other knowledge-based industries, knowledge can be viewed as a 'stock of expertise'. An organisation's stocks of expertise come from the flows in complex input–output systems. Knowledge flows in through hiring, training and purchase of capital goods. Some knowledge gets 'manufactured internally', through research, invention and culture building. Knowledge flows out through staff departures, imitated routines and sales of capital goods.

The organisation of the supply chains is an important characteristic of construction organisations. These supply chains exhibit a specific division of labour and institutionalised roles, such as engineering companies and architects, building contractors, and the manufacturers and suppliers of building materials or parts.

3.4.1 *Intra- and interorganisational knowledge sharing*

In the UK construction industry, there is now a steady increase in collaborative working practices, such as partnering, alliances, joint ventures, framework agreements and private finance initiative (PFI) projects. In addition, projects are growing in complexity and cost, and clients' demands and expectations are also increasing more than ever before. This presents a situation where organisations have to collaborate and share knowledge, skills and expertise in order to meet the needs of clients. However, organisations need to be mindful of both the communicative behaviours and practices associated with knowledge exchange and the 'knowledge paradox'. Organisations will have to be open to, formal and informal, information and knowledge flows from both networks and markets. At the same time, they must protect and preserve their intellectual capital and knowledge base because it is upon this latter point that survival depends. Many construction organisations have started to develop extranets to facilitate collaborative working on specific projects. However, external knowledge sharing poses greater risks than internal sharing – raising complex issues such as confidentiality, reliability and copyright. Matusik and Hill (1998) argued that 'more permeable firm boundaries provide for easier access to external knowledge but, at the same time, allow for more rapid dissemination of a firm's unique stock of knowledge outside its boundaries'. They further argued that 'grafting knowledge from the outside environment is not easy, a firm needs mechanisms to bring public knowledge in, to transmit this knowledge within the firm and to fuse the new knowledge with existing stocks of knowledge'.

Many innovation processes in the management and procurement of construction activities are becoming increasingly interactive, requiring simultaneous networking across multiple communities of practice such as professional groups, functional groups and business units (Egbu *et al.*, 1999). This networking involves communication and negotiation among different social communities with distinctive norms, cultural values and interest in the innovation process (Egbu *et al.*, 2000). This therefore means that the knowledge needed for innovation and competitive advantage is distributed within organisations and across organisational boundaries. Heterogeneous team organisations cut across these boundaries and are therefore important for realising the products of the knowledge economy. The movement of personnel could also be seen as a mechanism for distributing tacit knowledge and skills, or human capital, across space and time.

Construction professionals and organisations face economic imperatives when they conduct their knowledge work. More codification of knowledge, and more explicit repositories, enhances knowledge costs, efficiency of exploitation and transparency of sharing. However, Sternberg (1988) argued that much of the knowledge on which performance in real-world settings is based is tacit. Tacit knowledge, however, does not imply that such knowledge is completely inaccessible to conscious awareness, unspeakable or even unteachable; merely that it is not usually taught directly.

Explicit repositories of knowledge often hide the in-depth nature of knowledge work, which includes the articulation of collective tacit knowledge into a strategic collective expertise. It could be argued that the competitive advantage of professional firms in the construction industry lies in their ability to build communities of practice, relationships with their clients, and to increase their tacit and collective knowledge capital through intensive socialisation. Professional practices in construction may therefore take seriously their ability to shift from accumulation-driven KM towards more sense-making and tacit understanding-driven knowledge capital.

Network relationships and collaboration are key sources of knowledge building, language sharing and in the building of shared meaning. Organisations working together in networks such as supply chains are likely to spread and share best practices and the results of research and development. However, organisations would need to be mindful of adaptive efficiency and role boundary spanning when dealing with the different types of organisational co-operation and collaboration.

3.4.2 Communities of practice (CoPs)

An important concept in relation to building appropriate networks is that of a community of practice (CoP). Communities of practice (CoPs) are developed through the process of acting together, to create meaning through negotiation (Brown and Duguid, 1991; Lave and Wenger, 1991; Wenger, 1998). The negotiation of meaning has two complementary elements, 'participation' and 'reification'. From the perspective of a single CoP, 'participation' refers to the social interaction of members that keeps the CoP together, and 'reification' stands for the material manifestations of a diversity of cognitive activities by which members create meaning. An important requirement for sharing knowledge between people is therefore a presence of shared practice. Shared practices are vital for developing knowledge as they enable the flow of knowledge within the group.

The communication of knowledge in CoPs is possible and effective between people who, to some extent at least, share a system of meaning. In other words, they have an 'absorptive capacity' to understand one another.

The use of CoP is an effective mechanism for integrating strategic business units (SBUs) and functional areas such as human resources, planning, information technology, design, marketing and training. Since all professions and occupations are affected by a KM approach, it is essential that all professions, including support professions, move out of their 'functional silos'. Within a CoP, the issue of handling knowledge in heterogeneous and temporary groups is important. Such groups are characterised by performing tasks with a high degree of complexity and lack of formal structures that facilitate co-ordination and control. They often entail high-risk and high-stake outcomes and depend on an elaborated body of collective knowledge and diverse skills. Moreover, there is mutual dependency of the participating partners, which stems from a division of labour where each task is dependent on the other.

A recent survey of large construction organisations carried out at Loughborough University (Carrillo *et al.*, 2002) showed that CoPs are the most widely used technique for knowledge sharing. Large international construction organisations with a range of specialist skills tend to have the greatest need as well as resources to set up CoPs and to benefit significantly from them.

3.4.3 *Networking*

Networking is not only important for gaining information and knowledge; it is also necessary in order to overcome barriers that may prevent individuals or teams being able to understand each other's perspectives. Organisations rely on groups or networks of specialists because they can access a larger amount and more diverse, yet relevant, information and knowledge than individuals, and thus possess the potential for improved task performance. Over time a group or network may develop a collective understanding or shared mental model. The team can have a shared mental model of the task, knowledge about work, and a shared mental model of the team, and knowledge about group members. Experience of working with the same group allows individuals to develop an understanding of others' abilities, enabling them to develop strategies for solving problems. The strategies are not known solutions but are knowledge bases that are highly regarded for certain types of information. Previous interaction enables a shared mental model, which reduces the amount of 'small talk' required before they can communicate on the relevant issue. An understanding of the communication behaviours of others should help communicators calibrate their communication so that informative intention is understood. Communication behaviours within an organisation play a significant part in contributing to or detracting from an organisation's success. This again highlights the importance of absorptive capacity among those sharing and communicating professional knowledge.

The construction industry has also been made increasingly aware of knowledge sharing through networks. Examples of knowledge sharing networks are the Construction Best Practice Programme (CBPP), Construction Productivity Network (CPN) and Movement for Innovation (M4I). The Co-operative Network for Building Researchers (CNBR) is also a useful network. Following the Egan (1998) recommendations, a number of benchmarking clubs have also been created in construction to facilitate learning and sharing best practice. This is consistent with what Bennett (2000) referred to as the 'third way' in construction, that is, the need to balance co-operation with competition.

3.4.4 Communication

Communication is a key requirement for building a project's, or an organisation's, knowledge base (Egbu *et al.*, 2000). More specifically, it involves interaction between specialists, moving information more freely between those with information/knowledge and those who may find the information useful. Building and maintaining networks is important if individuals and organisations are going to widen and possibly strengthen their decision-making potential. Maintaining communication can be used to prevent interaction deteriorating, improve interaction or as a routine mechanism for keeping knowledge exchange fluid. Being aware of strategies for maintaining communication and consciously using communication strategies to strengthen and build networks will be important for the strategic management of knowledge.

A business environment that is conducive to information and knowledge exchange would seem to be one that has a supportive culture, with participants who are actively engaging in individual and project objectives. The participants who are active within the project value their own and their colleagues' participation, and their knowledge. Participants must feel that they can express their true thoughts on the subject, and these must be valued. Wherever hierarchical structures, social power or competition are a dominant factor, individuals may be less willing to voice their true thoughts and opinions.

A problem with business communication is that people cannot be forced to provide or accept knowledge. Communication of ideas, information and suggestions in a competitive environment may not be forthcoming. If power, position or status is used to demand information, the requestor may obtain the information he or she wants without any positive intellectual contribution from the individual. Specialist information, which manifests in the person's mind, is exchanged, in most cases, when the person chooses to disclose it. When it is not known whether someone has specific information or knowledge, he/she cannot be forced into divulging it. Tacit knowledge resides in the individual person's mind and there is little chance of others knowing it unless the person chooses to disclose it.

Organisations and projects are complex, with complexity increasing with the number of multidisciplinary specialists involved. The project environment brings with it considerable information processing needs. Individuals within organisations and projects will determine how much effort they are going to expend assisting the processing of information. Individuals may be selective with the amount of information disclosed, the person to whom they disclose it and the degree to which they attempt to get the other person to understand. An organisation or project environment that facilitates the release and exchange of information needs to support individuals, groups and networks, making strategic efforts to assist information and knowledge flow. Organisations need communication strategies to reduce barriers and support co-operation.

3.5   Creating and sustaining a knowledge culture

Managing knowledge assets is not easy. Creating and sustaining a knowledge culture where knowledge is valued and where knowledge creation, sharing and utilisation are a natural and an instinctive part of business processes is also not easy, but essential for meeting the challenges ahead.

The following issues are important in creating and sustaining a knowledge culture for the challenges associated with the knowledge economy:

- Support from top management and the presence of a strong 'knowledge champion'.
- Link to economic performance and strategy, underpinned by a coherent knowledge vision and leadership.
- Clear purpose and shared language and meaning of KM.
- 'Flexibility' in the lines of communications, allowing top-down, bottom-up and lateral communications within organisations. Multiple channels of knowledge transfers/dialogue with functional departments, interaction with clients/customers and suppliers.
- A risk-tolerant climate, where it is accepted that lessons could be learned through mistakes.
- A climate where people genuinely feel valued and people feel some form of 'ownership' or are involved with the KM initiatives in place.
- A climate where people feel secure in their jobs.
- Technical infrastructure (systems to obtain, organise, restructure, warehouse or memorise and distribute knowledge) such as intranet, internet, repositories, databases and video-conferencing.
- Organisational infrastructure (teams, relationships and networks) including face-to-face meetings, brainstorming sessions, apprenticeships, job rotation, coaching and mentoring, CoPs, quality circles, reports and project summaries, help desks and bulletin boards.
- A sharing culture where there is openness and willingness to share information, experience and knowledge across project teams and the organisation.

- Change in motivational practices (including performance management and team-based rewards).
- An environment that supports and promotes education and training.

For a corporate-wide programme, the involvement of the Chief Executive Officer or the Managing Director is important. At a regional and departmental level, the involvement of top management is also crucial. The strategy for planning the approach to managing knowledge assets for an organisation will usually be developed by a small team drawn from top or senior management, representing different aspects. Representation and buy-in from the wider organisation is important, so as to ease the implementation and rolling-out stages of the strategy. Similarly, having robust organisational infrastructure, flexible knowledge structures and positive motivational practices should promote knowledge sharing. Further discussions in creating a knowledge-sharing culture in construction project teams are provided in Chapter 12.

3.6   Conclusions

Conducting business in a knowledge economy brings opportunities and challenges. The opportunities include the possibility of increasing market share, improving productivity and profitability through innovation and the effective management of knowledge assets. The main challenges involve dealing with increased global competition; the changing levels and patterns of demands from clients, customers and the society; and the increasing pace and implications of change in information and communication technologies (ICT).

The construction industry is a knowledge-based industry. It is diverse, being made up of different organisations, consultants and professionals providing a range of services for clients, customers and the wider community. The activities of today's construction industry also demand an increased level of knowledge, skills and learning; and this relies on the formation of hierarchies of networks driven by the acceleration of the rate of change and rate of learning. The industry can be viewed as a 'stock of expertise'. These stocks of expertise come from the flows in complex input–output systems. Knowledge flows in through hiring, training, and purchase of capital goods, while some knowledge gets 'manufactured internally'. Knowledge flows out through staff departures and imitated routines.

Knowledge is the driving force of a knowledge-based industry. It is critical for effective action in the economy of the future and can bring critical competitive advantage. Knowledge leadership is vital for the industry. The question for leaders in the construction industry and for management is how to optimise within the old constructs that still exist. For example, if the construction industry is to build and maintain capability in a knowledge economy, it has to change its adversarial culture to a sharing culture. Furthermore, it has to learn from each project and transfer knowledge from projects to organisational bases. The industry will also need to invest in long-term relationships.

For the construction industry to deal successfully with the challenges of the knowledge economy, it has to deal effectively with its skills shortage. Creating and sustaining a knowledge culture (where knowledge is valued and where knowledge creation, sharing and utilisation are natural and instinctive parts of business processes) is essential for meeting the challenges of the knowledge economy. This calls for effective vision, leadership, coherent strategies and structures, and respect for people.

References

Al-Ghassani, A.M (2003) *Improving the structural design process: a knowledge management approach.* PhD Thesis, Loughborough University, UK.

Bennett, J. (1991) *International Construction Project Management: General Theory and Practice.* Butterworth-Heinemann, Oxford.

Bennett, J. (2000) *Construction – The Third Way, Managing Cooperation and Competition in Construction.* Butterworth-Heinemann, Oxford.

Brown, J.S. and Duguid, P. (1991) Organizational learning and communities of practice: towards a unified view of working, learning and innovation. *Organization Science*, **2**(1), 40–57.

Carrillo, P.M., Robinson, H.S., Al-Ghassani, A.M. and Anumba, C.J. (2002) *Survey of Knowledge Management in Construction.* KnowBiz Project, Technical Report. Department of Civil and Building Engineering, Loughborough University, UK.

Construction Research and Innovation Strategy Panel (CRISP, 2001) *Changing Skills Needs in the Construction Industry.* CRISP Consultancy Commission, London.

Construction Research and Innovation Strategy Panel (CRISP, 2002) *Culture and People in Construction – A Research Strategy.* CRISP Culture and People Task Group, London.

Davenport, T.H. (2000) *Knowledge Management Case Study – Knowledge Management at Ernst and Young, 1997.* http://www.bus.utexas.edu/%7 Edavenpot/e_y.htm.

den Hertog, P. and Bilderbeek, R. (1998) *Innovation in and through knowledge intensive business services in the Netherlands.* TNO-report STB/98/03, TNO/STB 1997. Centre for Technology and Policy Studies, Apddoorn, The Netherlands.

Department of Trade and Industry (DTI, 1998) *Our Competitive Future: Building the Knowledge Driven Economy.* DTI, London.

Department of Trade and Industry (DTI, 2002) *The state of the construction industry report.* Produced by the working group of the Consultative Committee for Construction Industry Statistics, which consists of DTI and industry members. DTI, London.

Egan, J. (1998) *Rethinking Construction: Report of the Construction Task Force on the Scope for Improving the Quality and Efficiency of the UK Construction Industry.* Department of the Environment, Transport and the Regions, London.

Egbu, C.O., Sturges, J. and Bates, M. (1999) Learning from knowledge management and trans-organisational innovations in diverse project management environments. *Proceedings of the 15th Annual Conference of the Association of Researchers in Construction Management (ARCOM)*, Liverpool John Moores University, 15–17 Sept, vol. 1, pp. 95–103.

Egbu, C.O., Sturges, J. and Gorse, C. (2000) Communication of knowledge for innovation within projects and across organisational boundaries. *Proceedings of Congress 2000: 15th International Project Management World Congress*, Royal Lancaster Hotel, London, 22–25 May, Session 5: project management at all levels.

Hansen, M.T., Nohria, N. and Tierney, T. (1999) What's your strategy for managing knowledge? *Harvard Business Review*, March–April, 106–117.

Lave, J. and Wenger, E. (1991) *Situated Learning: Legitimate Peripheral Participation.* Cambridge University Press, Cambridge.

Matusik, S.F. and Hill, C.W.L (1998) The utilization of contingent work, knowledge creation and competitive advantage. *Academy of Management Review*, **23**(4), 680–97.

Nonaka, K. and Takeuchi, H. (1995) *The Knowledge Creating Company: How Japanese Companies Create the Dynamics of Innovation.* Oxford University Press, Oxford.

Nonaka, I. and Toyoma, R. (2003) The knowledge creating theory revisited: knowledge creation as a synthesizing process. *Knowledge Management Research and Practice*, **1**, 2–10.

OECD (1996) *The Knowledge-Based Economy.* OECD, Paris.

Rethinking Construction (2000) *A commitment to people – our biggest asset.* A report from the Movement for Innovation's (M4I) Working Group on Respect for People, London.

Rethinking Construction (2002) *Respect for people – A framework for action.* A report of Rethinking Construction's Respect for People Working Group, London.

Robinson, H.S., Carrillo, P.M., Anumba, C.J. and Al-Ghassani, A.M. (2001) Knowledge management: towards an integrated strategy for construction project organisations. *Proceedings of the 4th European Project Management (PMI) Conference*, London, 6–7 June, paper published on CD.

Robinson, H.S., Carrillo, P.M., Anumba, C.J. and Al-Ghassani, A.M. (2002) Knowledge management for continuous improvement in project organizations. *Proceedings of the W65 (Organisation and Management of Construction) 10th International Symposium*, Ohio, 9–13 Sept, pp. 680–97.

Scottish Enterprise (1998) *The Clusters Approach: Powering Scotland's Economy into the 21st Century*, p. 8. Scottish Enterprise, Scotland.

Sheehan, T. (2000) Building on knowledge practices at Arup. *Knowledge Management Review*, **3**(5), 12–15.

Sternberg, R.J. (1988) *The Triarchic Mind. A New Theory of Human Intelligence.* Viking, New York.

Strategic Forum for Construction (2002) *Accelerating change*. A report by the Strategic Forum for Construction chaired by Sir John Egan, London.

Wenger, E. (1998) *Communities of Practice: Learning, Meaning and Identity*. Cambridge University Press, Cambridge.

Windrum, P., Flanagan, K. and Tomlinson, M. (1997) *Recent patterns of services innovation in the UK*. Report for TSER project 'SI4S', Policy Research in Engineering, Science and Technology, Manchester.

4 Strategies and Business Case for Knowledge Management

Tony Sheehan, Dominique Poole, Ian Lyttle
and Charles O. Egbu

4.1 Introduction

Today's construction industry demands results faster than ever – decisions must be made rapidly through electronic communication, placing considerable pressure on the individual. Construction professionals must be constantly aware of past experience, yet must also seek to incorporate an ever-growing pool of new ideas in order to innovate faster than the competition. Failure is not an option. Global competition ensures there are few second chances – it is an age of 'right first time' or 'never again'.

In the face of such challenges, knowledge management (KM) offers real potential to organisations seeking to be as effective as possible. Unfortunately, the field remains ill-defined and often misused by management consultants and technology sales people. This chapter attempts to remove some of the uncertainty from the field, and to outline examples of strategies and successful practices that have been used to good effect by construction organisations.

Following this Introduction, Section 4.2 explores the meaning of KM to construction organisations and the key areas necessary for the effective delivery of KM. Section 4.3 examines the issues associated with selecting an appropriate KM strategy, distinguishing between a tacit approach (focusing on people) and an explicit approach (focusing on capturing knowledge in documents, databases, intranets, etc.). Section 4.4 discusses the practical implications of adopting a knowledge strategy in terms of the role of people, processes and technology. Section 4.5 presents the business case for KM, highlighting on the one hand the business requirements and expected benefits from KM, and on the other the need to quantify the benefits. Section 4.6 concludes that KM in the future will become an integral part of business practices.

4.2 What does knowledge management mean to construction?

The challenge of managing knowledge in the construction industry is highlighted by the *Harvard Business Review* definition of KM:

'The way companies generate, communicate and leverage their intellectual assets.'

This definition highlights five key areas that are vital to delivery of effective KM:

(1)   *'The way'* highlights that KM must be customised to each organisation – there is no one right way, and construction organisations must develop approaches consistent with their values and objectives.

(2)   *'Generate'* implies a need to address recruitment of the best staff and constant training of those staff in order to maximise their knowledge.

(3)   *'Communicate'* reinforces that there is no point in having knowledge unless it is shared with other people throughout the organisation/project team or, indeed, industry.

(4)   *'Leverage'* reinforces the need to use knowledge on construction projects. Capturing knowledge or having knowledge present within an organisation is of little use unless it is actually applied.

(5)   *'Intellectual assets'* recognises that people and knowledge that have been captured are an asset. Organisations that can successfully exploit existing knowledge from inside and outside the firm (e.g. from trade associations) can gain a powerful source of competitive advantage.

4.3   What knowledge management strategy should be adopted?

For construction organisations, good KM practice requires knowledgeable people who are supported by integrated information and data sources in order to generate informed decision-making, as shown in Figure 4.1.

Consider an engineer faced with a dilemma of how to avoid alkali silica reaction in concrete. A web search will reveal little more than data – unstructured facts that do little to inform the decision-making process. If

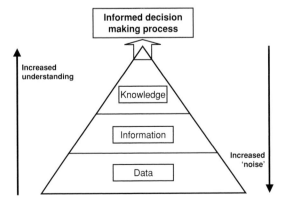

Figure 4.1   Knowledge support for decision-making

the engineer probes further, he may be able to find appropriate information within industry standards or concrete society guidance documents, knowledge that has been captured and shared. In some circumstances, however, the only solution to his problem will be found by talking to another person direct – searching out the increased understanding and clarity from the informed tacit knowledge of an expert in the field who can answer a detailed question quickly and easily.

For organisations seeking to make the first steps in their KM journey, it has been suggested that:

> 'Knowledge Management is never zero based: to make it work you need to recognise that you are already doing it.'
>
> (Birkenshaw, 2001)

Any organisation seeking to progress KM should relate initial projects to successful practices, whether relating to the tacit knowledge in people's heads or the explicit knowledge in documents and intranets. Organisations must also examine strategic priorities and assess how KM can support these – if it does not impact on bottom line, it is not worthwhile.

Construction organisations seeking to manage knowledge have essentially two options:

(1) Organisations could seek to capture construction knowledge in documents, databases, intranets, etc. This 'explicit knowledge' approach works well for standardised problems, but does little to enable exchange of new ideas.
(2) Alternatively, organisations could focus primarily on people and develop ways in which they can exchange their 'tacit knowledge' in order to facilitate innovation. This can, however, allow excessive duplication and wheel reinvention in some cases.

The appropriate balance of 'explicit' versus 'tacit' knowledge depends on each organisation's strategy. An organisation concentrating on dealing with standardised solutions to client problems will tend to adopt a strategy focusing on explicit knowledge. On the other hand, an organisation seeking to continuously innovate must address the far more challenging area of tacit KM, and would adapt its KM strategy accordingly. Any organisation will require elements of both approaches, and must integrate the two effectively.

Whilst some benefit may be achieved by introducing 'big bang' enterprise or large-scale 'knowledge management technology solutions', these routes to KM should be treated with caution. Technology solutions do not work 'off the shelf'; they require careful project management to realise potential benefits. In practice, most construction organisations will generally find it more appropriate to focus initially on the needs of their people or processes, then to develop appropriate tools in these areas to generate a firm foundation of knowledge sharing throughout the organisation.

4.4   Delivering knowledge management in practice

Most construction organisations will have some element of KM already in place. It is important, however, to explore the effectiveness of existing KM efforts, and to map out workable KM strategies for the future. In broad terms, many organisations have found it beneficial to explore KM in terms of people, process and technology.

The KM solutions to these three areas are neither instantaneous nor easy. Solutions must be customised to organisational strategy and will require the commitment of senior management and the buy-in of the whole organisation. Each area is considered in turn.

4.4.1   People and their role in knowledge management

People are known to be key to successful KM. The importance of addressing people aspects in any KM strategy is highlighted by such factors as:

- Some 80% of useful knowledge is tacit and cannot be written down; KM strategies must recognise and accept this, reflecting on how to mobilise appropriate tacit knowledge at the right place and time.
- The construction industry is characterised by a wealth of experiential knowledge, yet senior staff retire or leave organisations regularly, potentially taking tacit knowledge and a potential source of competitive advantage with them. Organisations must each decide how best to cope with this problem such that as much knowledge as possible is retained within organisational boundaries.
- Marshall and Sapsed (2000) highlight that
  - knowledge resides in the heads of senior engineers who have gained it through experience
  - many of these engineers see 'knowledge as power' and do not readily share
  - many experienced engineers are approaching retirement
  - there is limited knowledge transfer to more junior staff.

At its simplest level, the challenge is to know who knows, transferring knowledge of key staff both locally and, in larger firms, internationally.

A key challenge to construction organisations is to ensure that knowledge transfer exists effectively within teams, particularly from experienced staff to new graduates. The simplest and most effective approach is to develop a form of mentor/apprentice relationship, which has proved so essential to knowledge transfer in the craft side of the construction industry over time. This form of local knowledge transfer is possible at team level, but transferring knowledge between teams and indeed across organisations is a larger challenge.

Many construction organisations have sought to ensure that the knowledge of their people is made available at the right place and time. At Arup, they found that the size of the firm was making it increasingly difficult to *'know who knew'* the answer to a particular problem. In response, a web-based knowledge sharing tool was developed that now allows Arup to locate experts throughout 7000 staff in a matter of seconds. Each individual, from chairman to new recruit, can quickly and easily share their interests and expertise with the rest of the firm.

Similar approaches have been adopted at Taylor Woodrow, Thames Water and Mott McDonald. In each case, the aim is to understand what knowledge exists within teams, to facilitate the creation of powerful networks such that this knowledge is mobilised across or between similar teams, and have an appropriate culture such that people within the organisation are always seeking to learn from each other. Clearly, some of these issues cannot be achieved by technology alone, and issues of cultural change, mentoring, appraisal and reward will all have an impact.

The type of 'corporate yellow pages' developed will vary according to the type of firm within which it is rolled out. For hierarchical organisations, a strong degree of validation of skills will be required, which will limit the effectiveness of this kind of application. Within Arup, a far more organic organisational structure exists, and it was possible to allow the individual to 'volunteer' information, with minimal checking of content by any central source. Each individual has, in essence, their own web page where he or she is able to quickly and easily share interests and expertise with the rest of the firm.

The freedom of this approach has ensured a high level of completion and that content is constantly live and regularly updated. Careful content management and maintenance procedures were found to be important to maintain currency of content, together with links to the appraisal process to ensure completion.

In addition to knowing who knows, however, a key challenge is encouraging people to use knowledge – creating a culture of knowledge sharing and use where:

- it is acceptable to ask for help
- it is reasonable to make mistakes
- it is possible to share lessons in a culture of continuous improvement
- people actively seek and apply new learning.

Culture has been widely recognised as a key factor in successful knowledge-sharing initiatives. Allday (1997) highlights that 'Having an open and participative culture which values the skills and contributions of employees at all levels is critical'. Culture was also recognised as a major barrier to the creation of knowledge-based organisations by 80% of respondents to a survey by Chase (1997).

Construction organisations have responded to the challenge of culture in several ways. At Arup, they were fortunate to have a strong culture of knowledge sharing which can be traced back to the founder of the firm in the 1940s. Arup has also adopted a philosophy of encouraging free time for knowledge sharing and innovative projects – an approach used to encourage appropriate behaviours for KM. Other firms such as Halcrow have used performance coaching to help cultivate appropriate knowledge behaviours.

Many organisations have recognised that the success of KM efforts comes down to people and their behaviours. To achieve appropriate results, many organisations may have to consider targeted cultural change programmes. Once appropriate behaviours have been defined, however, it is important to reinforce the key role of these practices by checking through discussion at appraisal and using KM as a key factor in performance appraisals.

4.4.2 Processes and their role in knowledge management

Given the increasing drive towards lean construction in recent years, many construction organisations have tried to embed good knowledge-sharing practices into daily activities by making them part of core business processes. The extent to which a formal process-centric approach can be useful for KM again depends to a great extent on the organisational strategy and focus. For process-centred firms, it may well be possible to 'manage knowledge into core processes' to a far greater extent than those firms seeking to continuously deliver innovation and creativity. The challenge is to capture as much useful content as possible, and then to apply it in projects at the right place and time.

Approaches that have been successfully applied include:

- embedding good KM practices into processes to help transfer knowledge across the business
- creating a network of virtual knowledge leaders to cascade good KM practices to the coalface
- developing a network of divisional knowledge activists to help 'spread the word'
- aligning KM with QA following the ISO 2000 approach such that KM can become embedded.

Other key process tools that can be used by firms seeking to manage knowledge include:

- best practice documents
- project reviews and organisational learning
- communities of practice.

Creation of a reliable, validated source of best practice is a simple, effective way of guiding construction firms to reduce wheel reinvention and apply best practice wherever possible. These documents take many forms.

At Arup, the feedback note system is written by the firm's experts who share best practices, watch-its and industry trends; and at EC Harris, the knowledge manager helps to produce high-quality knowledge packs on key topics of relevance to the business. A study on tall buildings, for example, provides easily accessible nuggets of knowledge relating to cost and construction issues, reducing wheel reinvention and improving effectiveness.

In all of these cases, the key is to first identify the firm's experts, then empower them with the ability to write these documents, then encourage them to share the documents throughout the firm, either electronically or in paper form.

Reviews are a powerful approach to transferring knowledge between projects, providing an excellent opportunity for the introduction of tacit and explicit knowledge at the right place and time. Many organisations have made the link between knowledge transfer and learning, and have sought to mimic the BP approach of learning before, during and after projects, at each stage pausing to reflect on the effectiveness of current approaches and on the viability of any alternative approaches to the problem at hand.

Construction projects are inherently complex and challenging, offering frequent opportunities for learning from daily activities. Reflecting on past experiences can enable project-based competencies, leading to sustained competitive advantage. There are several methods that can be adopted to achieve project learning, but all have the common aim of enhancing understanding of key project experiences. Lessons learned are acquired by acknowledging the successes and failures and applying this insight to improve future projects.

Project learning is complementary to an overall KM strategy. Approaches such as after-action reviews and learning histories are applied to capture tacit and team knowledge. Facilitating lessons learned and the successful dissemination and application of that knowledge to future projects should help to prevent reinventing the wheel each time a similar problem is encountered. Successful learning can potentially offer organisations many benefits, including greater predictability, lower defects, greater efficiency and reduced project risk.

Each individual involved in a project will possess specific experiences and it is this collective tacit and team knowledge that learning approaches seek to capture and exploit. Ideally the knowledge-harvesting approach should be integrated into the project processes as far as possible to avoid the perception of project learning as a separate task. Invariably project teams disperse after project completion and quickly begin their next job. This factor, combined with a lack of incentive or time, creates little opportunity for reflection of experiences. This highlights the importance of establishing systematic learning processes through reviews that are built into the project process.

It is particularly important, however, to look back at projects. It is only with the benefit of hindsight that it is really possible to reflect on the true consequences of an action within a project, seeking to explore, for example:

- Did a new material perform well?
- Was a decision appropriate?
- What were the significant causes of success or failure?

The answers to these questions can provide a rich source of learning for both the project team and the construction organisation. Many firms, however, have found it beneficial to structure project close-out reports in a way that facilitates exchange between project teams. A variety of established methods can be applied to harvest project knowledge, including after-action reviews and learning histories.

- *After-action reviews*
  The after-action review (AAR) was originally developed as an approach to learning by the US Army in the 1970s. It was designed to facilitate day-to-day learning from combat training exercises. The approach is a discussion of an event that enables the individuals to learn the reasons for success or failure. Key questions are asked that intend to investigate what happened, why it happened, what went well, what could have been improved and the lessons learned.

- *Learning histories*
  Learning histories (originally developed by the Massachusetts Institute of Technology (MIT)) is a method of documenting and analysing project experiences and essentially presents a collection of individuals' experiences through narrative. It is best suited to complex projects involving a variety of individuals and is a flexible and practical way of gathering project information once the project team has become dispersed. The project knowledge is primarily captured through a series of interviews with members of the project team and can be supplemented by project records.
  The narrative document is presented in a two-column format organised in segments forming a series of short stories. These segments may represent the different stages in the project process such as appointment, briefing, scheme design, etc. In the right-hand column key experiences and events that occurred throughout the course of the project are described through quotations of the interviewees. The left-hand column contains the analysis and commentary that distils the lessons learned. The whole document is divided into segments which recount a particular episode and is headed by a short prologue addressing particular problems or dilemmas.

In addition to formal, structured reviews of this form, it is also possible to carry out more informal knowledge exchange across projects through, for example, communities of practice. These can be used to good effect to exchange knowledge across the organisation and provide a means of semi-informal review across the firm. At Thames Water, project experiences are routinely captured in the areas of both business learning and project innovation. Experiences of both Thames Water and their alliance partners are shared throughout the firm via the intranet, with a series of communities of practice helping to review the validity of knowledge and apply it successfully to resolve problems.

One of the most valuable aspects of communities is their ability to host and actively participate in discussion fora, helping people to provide rapid access to answers to questions as and when they occur on real projects. One of the challenges of communities is, however, that they are 'easy to kill, but difficult to manage'.

Communities thrive because people want to play a part. They survive best in empowered organisations where people are ready and willing to contribute. In contrast, they may suffer in more formal, hierarchical organisations where people perceive communities as less business critical than project teams, and overmanagement may result.

4.4.3   Technology and its role in knowledge management

One of the greatest problems facing any construction organisation seeking to embrace KM technology at present is gaining a realistic understanding of the field. Many technology providers are eager to sell off-the-shelf solutions, but comparatively little exists in the way of objective guidance on which technologies are appropriate for individual company needs. Any construction organisation seeking technology solutions to KM must identify clear areas where KM could be enabled through technology.

KM technologies relevant to the construction sector can be divided into four broad categories:

(1) People-supporting (for example) profiling, mapping of skills and corporate yellow pages applications.
(2) Projects-supporting collaborative working, document management, reviews and archiving.
(3) Organisations-supporting cross-project and interdivisional working, communities of practice and idea generation.
(4) Industries-supporting extranet communities that unite disciplines across organisational boundaries.

Inherent within these categories is the need for integrating technologies, pulling together the separate knowledge areas in order to deliver added value for the organisation as a whole.

The appropriate KM technology to adopt for any construction organisation will vary according to each organisation's emphasis on people and process management. Having identified requirements, it is important to be selective in the KM technology market, negotiating sensible arrangements with best-of-breed products to satisfy organisational needs.

The backbone of almost all KM technology efforts at present for large firms is likely to be an intranet, with key components selected according to the functionality required from the areas above. People-based KM tools are geared to improving the understanding of people, their skills and their ability to apply these skills to the right place and time, whereas process-based tools such as extranets are very much more geared to ensuring the right piece of explicit knowledge is applied to projects at the right place and time. Project collaboration tools have much to offer in this space, as they facilitate effective communication across project teams. The underlying workflow within such systems can be used to extract key documents from projects and share them across the enterprise, and can also search past project databases and 'push' previous experiences to teams at the beginning of projects.

The value of a static knowledge repository has been questioned in recent years, and there is an increasing recognition for additional, informal review across project boundaries. It is possible to supplement the structured nature of project collaboration tools with less formal 'peer review' tools that facilitate communication within communities of practice.

For the organisation, it is necessary to create links across the various areas, providing a simple, user-friendly point of access to organisational knowledge, but tailoring content to individual preferences. Organisational KM tools sit to some extent between people and process-centred KM tools, forging 'people to content' and 'people to people' links as required to increase the effectiveness of knowledge sharing throughout the organisation. Organisational knowledge systems will ideally be driven by client need, constantly defining and refining the valuable knowledge within the organisation, but will also relate to people needs, reflecting current and desired competency profiles to achieve desired performance.

If organisational KM technologies are to be effective, they must be simple to use and must complement, not conflict with, existing internal tools. The latest generation of search and retrieval engines falls into this category, providing ever more advanced algorithms to enable pattern matching and knowledge retrieval across the enterprise. Learning management systems are also vital, providing theoretical access to learning on demand (Maurer, 2001).

The effectiveness of organisational KM tools is, however, frequently limited by the quality of content retrieved. Organisations must ensure effective content management procedures in order to ensure that any content retrieved is current and valid; failure to do so will immediately undermine the effectiveness of organisational KM solutions (Harrington, 2001).

For the construction industry as a whole, considerable work is required to integrate the diverse efforts in the KM technology space. Currently, the global construction industry has many project extranet systems, many conflicting technical knowledge bases and a lack of focus in terms of next steps. This is, however, starting to change. Increasingly, global standards institutions are making content available online, in order to enable effective mobilisation of industry standard knowledge around the construction world. Professional institutions in the UK are supplementing the British Standards Institution (BSI) by collaborating in order to provide a shared knowledge resource for construction professionals seeking best practice data (see construction.co.uk). Similarly project extranet providers are increasingly realising that collaboration between project extranet systems is as essential as collaboration within it – the need for compatibility and common standards is slowly, but surely being addressed.

Having reflected on the strategies for KM in construction organisation, the next part of this chapter will reflect on making a business case for KM.

4.5   A business case for knowledge management

In the current business climate, there is a growing need to clearly identify, evaluate and measure the concrete impact of knowledge initiatives or projects on business and organisational performance. One of the core issues of KM is to place knowledge under management remit in order to get value from it and to realise intellectual capital. The intellectual capital can be regarded as a major determiner of the difference between a company's 'book price' and the total value of its stock. For a large successful company, this difference can be considerable, representing the divergence between the way the organisation is seen by accountants and the way it is seen by the market. As an example, there is a great difference between the book price and share value of recently launched biotechnology companies, whose market value is clearly based on their knowledge assets, rather than traditional capital. However, it would seem that while the world of business is experienced in managing physical and financial capital, organisations have some difficulties in finding ways of exploiting knowledge and evaluating the impact of such knowledge on business practices and performance.

Justifying the benefits of KM initiatives is therefore vital. However, it is not easy. Since KM can represent a significant investment for an organisation, it is arguable that KM should be applied to business areas that will yield the best and most value.

Although there is much said and written about the benefits of KM, not as much is written about making a case for KM or measuring the benefits of KM initiatives. One of the reasons for this is the intangibility of some of the benefits of KM.

4.5.1   A business case: what the people within the business care about

From a KM perspective, a business case is often seen as a document that presents a comprehensive view of the knowledge initiative(s) or project(s) and provides the financial justification for implementation. It is a critical component of the initiative. The business case can therefore be used to communicate the project to others and to establish a method for measuring and for receiving funding approval for the initiative or project. A good business case should help to answer some of the following important questions:

- Why is the organisation/unit embarking upon the knowledge initiative/project in the first place?
- What is the initiative or project about?
- What is the 'solution' of the organisation/unit to the 'business problem', which the knowledge initiative seeks to put right?
- In what ways does the 'solution' address the business issues/goals?
- What is the cost (estimated/actual; also opportunity cost) of the initiative/project?
- In what concrete ways will the organisation/unit benefit from the initiative?
- What is the return on investment (ROI), if applicable and if appropriate to the organisation? What is the 'pay back' period?
- What are the risks of doing and not doing the project/embarking on the initiative, and are there alternatives?
- How will the 'success' and 'failure' be measured?

To demonstrate a business benefit, knowledge must be equated with something that the people within a business care about. It is important for an organisation to discover its own 'hot button' (specific KM goals). For example, an organisation may want to excel in one or two key areas, such as technical excellence, customer intimacy, superior quality, production efficiency, faster time to deliver project outcomes, staying on top of the competition, or cost management. Having identified the key area(s) of focus, attention should then be focused on these in terms of evaluating and demonstrating a business benefit. Again, depending on an organisation's specific KM goals, it may choose to target, evaluate and measure expense reduction, productivity gains or leverage improvement, or focus on the soft benefits of KM to mention but a few.

4.5.2 The business case and the value of knowledge: the quantification paradox

The value of knowledge can be considered in different ways. However, two main ways in which it can be looked at are through what could be considered as the 'micro viewpoint' and the 'macro viewpoint'. By considering the micro perspective, such issues as how can the impact of single knowledge initiatives/projects be assessed and quantified come to the fore. Such a knowledge initiative could include the roll-out of knowledge bases and idea generation systems as well as 'soft' interventions such as communities of practice, quality circles and story-telling.

On the other hand, the macro viewpoint could involve quantifying the intangible assets of an organisation by using such tools as the balance score-card, score boards, indexes and 'navigators'. According to Karl-Erik Sveiby (1997), the concept of intangible assets attempts to capture the value of human capital, competencies, customer relationships, employee collaboration or diversity in an organisation. Based on these concepts, tools such as Skandia Navigator have been created to serve as strategic and monitoring devices.

A case could be offered as to how the two perspectives (micro and macro) relate to each other. It is arguable that the main benefit of the macro perspective or approaches is that they allow an organisation to consider performance indicators that are purely financial. This is on the premise that the ultimate performance of an organisation is down to its intangible asset. On the other hand, most financial indicators essentially refer to past performance and therefore reflect outcomes as opposed to the value-generating drivers in an organisation. The crux of the matter and the main challenge in all these is to establish a direct causal link between concrete knowledge initiatives and their impact on business performance. This difficulty should not and, indeed, does not prevent some practitioners and organisations from evaluating the impact of their KM initiative, although this is normally not through 'measurement' in its strictest sense. It is often the case that support from senior management for KM initiatives requires some evidence of success. This has led to varying pragmatic ways by which the impact of knowledge initiatives can be demonstrated. These include anecdotal and case study evidence, feedback from users/participants, user survey and indirect measures (e.g. system usage).

This raises another important issue in the debate. This is the issue of the measurement paradox. Both quantitative and qualitative approaches have a role to play in evaluating the impact of knowledge initiatives. In certain contexts, quantitative measures may be actually limiting. As an example, system usage is easy to measure, but this does not guarantee that this will actually lead to individual, business or organisational performance. Such a measure can be seen as too 'indirect'. However, the impact of qualitative or 'soft' measures such as quality circles or communities of practice will be even more difficult to assess on the basis of their impact in terms of time saving, 'amount' of learning or financial value added. It could therefore be argued that the apparent precision of quantitative measures is offset by the fact that they often do not really measure what they are supposed to measure. In some cases, therefore, anecdotal evidence and case studies may be more appropriate or useful.

For some specific knowledge initiative or project, the ROI case can be proven as their outputs can be directly related to financial gains. It is, however, important to note that ROI is only able to capture part of the impact of the knowledge initiative. The reason for this is not too difficult to comprehend. Knowledge initiatives always have unintended consequences or effects (negative or positive) which cannot be easily captured as 'financial return'. These effects can therefore potentially undermine the validity of a good ROI calculation. Models based on ROI generally seem to have more validity when knowledge initiatives focus on efficiency and productivity concerns and less so when knowledge initiatives focus more on intangible assets, such as cross-project learning or competency development. In the latter case, a good 'theory' may be better ammunition to use in convincing senior management to commit resources to KM.

In certain instances, a good 'theory' can be supported by quantitative measures if used carefully and sensibly. As an example, Sveiby's collaboration climate index (CCI) (Sveiby and Simons, 2002) is a measure that can support the business case for improving collaborative relationships between employees. The main thrust of the CCI is that a good collaborative climate enhances knowledge sharing and, by implication, the development of intellectual capital. Although the index does not necessarily tell you how to actually improve the collaborative climate, nonetheless, the CCI is a useful heuristic device for understanding the factors that are important for collaboration and can therefore inform concrete knowledge initiatives.

Detailed aspects of KM performance measurement are covered in Chapter 9 of this book. However, from the above discussion, it becomes clear that an organisation's specific KM goals and the strategy it chooses to adopt are fundamental to the way it approaches the business case. It is this strategy for KM that determines how an organisation is able to fully create and exploit the knowledge available to it.

4.6   The future

As we look to the future, Peter Drucker (Drucker, 2001) has recently suggested that:

- Knowledge will be a critical resource.
- Knowledge will transfer more effortlessly than money via the internet.
- The knowledge society will be incredibly competitive.
- IT will allow knowledge to spread near-instantly.

In the face of these changes, the construction industry will have to demonstrate exemplary KM, applying the best tacit and explicit knowledge to problems wherever they occur.

In the future, KM will become the norm, embedded in organisational practice rather than regarded as a new field. The early signs of this trend are already in place, with KM standards available in the UK and Australia, and the subject of discussion in the US, Asia and through ISO.

Achieving successful KM will not be easy; organisations will have to simplify access to knowledge, both to increase the effectiveness of knowledge sharing within firms, but also to facilitate collaboration across supply chains. An appropriate culture will also be key; whilst it is possible to 'force' people into completing knowledge profiles or to push content to them, this approach is not sustainable. In practice, the future success of any KM system relies on creation of a culture to 'enthuse your employees with a desire for knowledge'.

Individuals must seek knowledge and must perceive a benefit for themselves as well as to the organisation if successful KM is to be achieved. Creation, and indeed maintenance, of such a culture is far from easy, but is essential to the success of any KM effort.

References

Allday, D. (1997) *Spinning Straw into Gold: Managing Intellectual Capital Effectively.* Institute of Management, London.

Birkenshaw, J. (2001) Why is knowledge management so difficult? *Business Strategy Review,* **2**(1), 11–18.

Chase, R.L. (1997) The knowledge-based organization: an international survey. *Journal of Knowledge Management,* **1**(1), 38–49, 83–92.

Drucker, P. (2001) *The Next Society.* Economist.com under the subheading 'Knowledge is all'. http://www.economist.com/surveys/displaystory.cfm?story_id=770819

Harrington, A. (2001) Sweet content. *Knowledge Management Magazine,* June, 14–16.

Marshall, N. and Sapsed, J. (2000) The limits of disembodied knowledge: challenges of inter-project learning in the production of complex products and systems. In: *Proceedings of Knowledge Management: Concepts and Controversies Conference,* University of Warwick, Coventry.

Maurer, H. (2001) Turn your knowledge into competitive advantage. *Knowledge Management Magazine,* May, 32–34.

Sveiby, K. (1997) The motivational impact of utility analysis and HR measurement. *Journal of Human Resource Costing and Accounting,* **2**(1).

Sveiby, K. and Simons, R. (2002) Collaborative climate and effectiveness of knowledge work – an empirical study. *Journal of Knowledge Management,* **6**(5), 420–33.

5 Organisational Readiness for Knowledge Management

Carys E. Siemieniuch and Murray A. Sinclair

5.1 Introduction

As part of the work for the CLEVER project (a two-year project funded by the UK Government's Engineering and Physical Sciences Research Council (GR/M72890) concerned with the development of a cross-sectoral process framework for knowledge management (KM)), it was realised that organisations need to be able to evaluate 'organisational readiness' for KM prior to considering how best to implement appropriate KM processes. This chapter addresses this issue, and is neatly summarised in the saying, 'If you would plant roses in the desert, first make sure the ground is wet'.

The original functional context for the CLEVER project was project management, since most business sectors in manufacturing and construction tend to adopt a project approach in order to carry out a range of vital operational and innovative activities. The leverage of project-based activities on company performance is becoming increasingly important, especially in such areas as new product introduction. However, the management of project knowledge (i.e. its collection, propagation, reuse and maintenance) is generally accepted as being open to considerable improvement, both within companies and between companies in the supply chain (Siemieniuch and Sinclair, 1993, 1999). Even where good practice can be identified, it is usually only disseminated within that industrial sector, with almost no learning across sectoral boundaries. Knowledge is generated within one project and then buried in unread reports and arcane filing systems, or lost because people move on (Markus, 2001). Failure to transfer this knowledge leads to wasted activity and impaired project performance. This problem is exacerbated by the increasing tendency to implement projects by temporary, 'virtual' organisational groupings where the responsibility for learning can become diffused across the supply chain and by the rapidly increasing complexities of supply chains (Gregg, 1996). For example, project teams in the construction industry are virtual organisations involving individuals in several firms who are disbanded at the end of a project and may never work together on another project.

Other aspects of CLEVER concerning the reasons for the existence of the project are reported in Chapter 10 and elsewhere (Leseure and Brookes,

2000; Kamara *et al.*, 2001); also the rationale of the tool produced to assist companies, and evaluations of the tool for use in solving KM issues in collaborating companies (Siemieniuch and Sinclair, 2002).

This chapter is concerned with the 'organisational readiness' issue; ensuring that the organisation is ready to adopt the philosophy of knowledge lifecycle management (KLM) as a fundamental *modus vivendi*. The chapter discusses the issues to be addressed in preparing organisations for the introduction of KLM processes. If one views organisations as 'knowledge engines' providing value to customers, then processes for KLM are fundamental for the organisation's survival. However, as with all capability, the organisation must be prepared for its introduction if it is to make profitable use of this new capability. The chapter includes findings from a number of other studies carried out by the authors over the last decade in collaborative projects concerned with co-operative working, KLM and manufacturing supply chains.

Following the Introduction, Section 5.2 discusses the importance of KLM, setting the scene for Section 5.3 which discusses preparing the organisation for KLM. This section outlines both strategic and tactical issues to be addressed – a devolved structure, IT-based tools, an open communications infrastructure, 'best practice' business processes, metrication, knowledge capture and a culture of trust – and outlines some processes necessary to establish the organisational context for KLM. Section 5.4 brings the chapter to a close, with some general conclusions and acknowledgements.

5.2   The importance of knowledge lifecycle management (KLM)

We include the word 'lifecycle' within KM because it is evident that organisational knowledge does indeed have a lifecycle; it is discovered, captured, utilised and, eventually, retired (or lost) rather than killed. There is a vast literature on KM of which perhaps the two most influential sources are Stewart (1997) and Davenport and Prusak (1998). Although there is increasing recognition of the role of humans in organisational configurations of knowledge within this business perspective, there is, as yet, no agreed ontology for analytic discussion, simulation and implementation of new forms of KM for learning organisations. This was the academic context for the CLEVER project.

From an industrial point of view, there is no doubt of the perceived importance of KLM. The close attention paid by many companies to the publicised initiatives by Skandia, the OSTEMS report (OSTEMS, 1998), the strong attendance of companies at workshops on KM, and initiatives in the UK by companies such as BP, British Airways, Unipart, Anglian Water and so on all demonstrate the business significance of KM. Perhaps the most potent reason for any organisation to adopt a KLM perspective of construction is when it considers its future in a changing environment. As one perceptive senior manager responsible for computer-aided engineering in an automotive company remarked, in 1997:

'In 30 years' time, we will be designing products we don't know, incorporating materials which haven't been invented, made in processes yet to be defined, by people we have not yet recruited. Under these circumstances, all we can carry forward is our knowledge, and our knowledge of how to improve our knowledge.'

It would seem sensible, therefore, to make a paradigm shift to viewing companies as 'knowledge engines', acquiring capabilities in the form of knowledge and skills, and utilising these in knowledge-intensive processes to deliver value-for-money products. There are some advantages in such a viewpoint. In brief, these are:

- It focuses management attention on the prime, essential assets of the organisation.
- It engenders a focus on strategic considerations.
- It presents a coherent and cohesive understanding of the enterprise's assets and capabilities.
- It is a precursor for a thorough understanding of process capabilities within the enterprise.
- It is the organisation's knowledge, rather than its physical assets, that differentiates it from its competitors, and this comes to the forefront of managerial consideration.
- Necessarily, a focus on knowledge implies a focus on good corporate governance, 'good management practice' (GMP) and on legal issues such as intellectual property rights, patents and so on.

It follows, therefore, that the processes that a company puts in place for KLM are key processes, requiring as much attention from the Board as the financial results. Furthermore, from a long-term perspective (and paraphrasing Crosby's famous dictum with regard to quality), 'effective application of organisational learning means that KLM activity does not represent an added cost; on the contrary, it is free' (Crosby, 1979). However, while there are clearly great benefits to be gained from KLM, there is still the issue of realigning the organisation to make the best use of this approach. In this respect, Denton (1998), Haywood and Barrar (1998), Snowden (1998) and Despres and Chauvel (1999) collectively make the point that organisations and individuals need to exhibit certain characteristics in order to learn: this is important since organisations cannot expect to implement KLM processes successfully (defined as achieving all their goals) in an environment that is not conducive to their execution. These characteristics include the following:

- A learning strategy – learning becomes a habit.
- A flexible structure – reduces bureaucracy and encourages cross-functional co-operation.
- Blame-free culture – encourages experimentation.
- Shared vision – establishes an overarching goal to help people pull in the same direction.
- Knowledge creation and transfer – leads to new products/processes and dissemination.
- Teamworking – helps combine existing knowledge and creates new knowledge.

The requirement for some of these characteristics is echoed by Leonard-Barton (1995) in her knowledge-inhibiting factors:

- limited problem-solving capability
- sterile implementation and inability to innovate
- limited experimentation in operations
- screening out of new knowledge.

Haywood and Barrar (1998), in a very useful contribution, raise issues to do with the capture and sharing of best practice and knowledge within teams, especially for small and medium-sized companies (SMEs) who may not have the resources to support their processes and, furthermore, may be forced by their larger partners in supply chains to adopt formats other than those they would like. It is these issues that are addressed by organisational readiness; how one prepares for KLM, with the intention of avoiding the usual, attendant problems associated with major organisational change. It is fairly well accepted in the literature above that there are seven bases on which organisational effectiveness depends:

(1)   A devolved organisational architecture (tasks, roles, jobs, teams, business units, support groups, head office) that facilitates the achievement of the organisation's goals – satisfaction of stakeholders, growth, flexibility and agility, etc.
(2)   IT-based tools for supporting business decisions, at all levels of the enterprise. Two of the biggest problems here are (1) projecting the future competitive environment (competitors' activities, changes in customers' perceptions, the changing industrial emphasis from products to services, changes in global politics and legislation, the effects of e-business, etc.) and (2) exploring the optimum configuration of the enterprise's resources in order to address this future environment in the most efficient manner (both the disposition of resources and their subsequent effective utilisation).

(3) An appropriate, homogeneous, modular, open IT and telecommunications infrastructure serving the whole extended enterprise.

(4) Revised, 'current-best', business processes that have been designed from a stakeholder perspective (customers, process user, managers, etc.).

(5) Process measurement, to establish current performance, to carry out bench-marking of processes, to set stretch targets and to identify opportunities for improvements.

(6) Efficient capture and utilisation of knowledge, both within the enterprise and from its environment. Three classes of knowledge are critical for the organisation's future, each of which comprises both formal and tacit knowledge. These are: (1) technological knowledge, covering products and processes (including patents, best-practice processes, etc.); (2) organisational knowledge, including knowledge about the organisation and its operations, the customer and the whole supply chain; and (3) network knowledge, which is inherent in the alliances and relationships that exist between the entities within the organisation and its networks, including suppliers, subcontractors, clients, consultants, universities and the like. This third class of knowledge constitutes a major source from which an organisation gains its flexibility, agility and future capability.

(7) The development and maintenance of a culture that supports organisational change and growth (Schein, 1996). There has been a wealth of literature written about all aspects of culture (e.g. Hofstede (1984, 1991), Hampden-Turner and Trompenaars (1994) and Trompenaars (1994) on national cultures; Handy (1976, 1989, 1994), Kotter and Heskett (1992) and Helmreich and Merritt (1998) on organisational culture; Levinson and Asahi (1995) on interorganisational learning; and Meier (1999) on professional and occupational culture). Organisational culture is more amenable to influence than professional or national culture and it is organisational culture that essentially channels the effects of the other two cultures into standard working practices. It is now widely accepted that while all seven bases are necessary, it is the cultural issues that are the most important and that can make or break any proposed change in an organisation (e.g. Revans, 1982; Eason, 1988; Huczynski and Buchanan, 1991; Lund *et al.*, 1993; Badham *et al.*, 2000).

This chapter addresses most specifically the sixth one of these bases (i.e. capture and utilisation of knowledge) in the sense that to erect a pillar on this base an organisation needs to have guidance on how to prepare itself for KLM and a process that will allow it to put itself in a position from which it will be able to execute KLM successfully. The remainder of the chapter attempts to provide some guidance on how to achieve this. It should also be noted that preparing an organisational context for KLM also provides foundations to support the other pillars (e.g. culture of change and growth, infrastructure, processes, organisational architecture, etc.).

5.3 Preparing the organisational context for knowledge lifecycle management

It appears that preparing the context must take place on a broad front. Since organisations always occupy some integral position in a supply (or value) chain, it would be highly advantageous if the whole value chain could move in concert towards a position where effective KLM is enabled both within and between organisations in the chain. However, unless such a chain has one highly dominant partner who is able to enforce compliance (as can occur in the automotive sector), this is unlikely to happen, and partners in the value chain will have to make their own provision, albeit within an agreed set of common goals and agreed standards. Issues and critical steps are described below, but it is emphasised that these formal steps apply mainly to large and medium-sized organisations; small organisations have a much more fluid organisational structure and can be less regimented in their approach.

From the perspective of the extended enterprise, a company should exhibit certain characteristics in its behaviour in the supply chain (Siemieniuch and Sinclair, 1999, 2000):

- transparency about goals, problems and ways of working
- willingness to share any benefits accruing from improvements or windfalls
- respect for commercial confidentiality
- recognition of personal relationships, built up over time
- recognition of the 'favour bank', for overcoming unexpected problems in the supply chain
- speedy, efficient, and correct execution of promises.

Therefore, the following human and organisational issues must be addressed:

- Organisational design must reflect the firm's role in its supply chain. Typically, this is interpreted as process-based 'business groups', etc.
- People who commit the firm to promises must be empowered and resourced to execute them. Good internal communications are necessary if the firm's responses are to be appropriate and coherent.
- The people executing the promises must have appropriate knowledge, skills and expertise, provided by formal training, learning-by-doing and by the other, formal, KLM processes put in place, as discussed below.

- Processes and procedures involved in the execution of promises must be clear (i.e. both the 'normal' operational procedures and established ways to get round problems occurring in these processes).
- While the people involved must be accountable for their errors, there must be appropriate, sensitive treatment of these errors by those in more senior positions.

Hence, the first steps within a supply constellation should be as follows:

- Identify common goals, policies and metrics across the e-supply network for KLM (Siemieniuch and Sinclair, 1999, 2000). These goals, policies and metrics must cover the four levels of interaction that must occur in e-supply chains:
  - transactional level – information about daily events must be communicated
  - operational level – there must be provision to co-ordinate and control the transactions (who meets, when and why)
  - policy execution level – negotiate targets, agree operational procedures, etc.
  - strategic level – define the role and level of participation in the supply chain, discuss market information and set other policy issues (e.g. define the type and scope of the contracts between companies).
- Move to partnership sourcing principles in the supply chain (Towill, 1996; Craig, 1998; MOD, 1999; Nadler and Tushman, 1999; Siemieniuch *et al.*, 1999; Siemieniuch and Sinclair, 2000). This will enable better information sharing to occur (including the use of common measures), it will widen the window of opportunity for improving the performance of processes (Rycroft and Kash, 1999) and it will allow more accurate assessment of the full, end-to-end process along the supply chain. For example, this may reveal that changes at other, external points in the supply chain will be of more benefit to the organisation than local changes.

Within such a framework, it then becomes possible for individual companies within a supply chain to prepare themselves for KLM at their own discretion. Figure 5.1 depicts some of the activities, and the dependencies between them, that need to be addressed before an organisation can begin to implement specific KLM processes (e.g. obtain or capture knowledge, locate and access knowledge, propagate knowledge, transfer knowledge, maintain or modify knowledge). The component parts of Figure 5.1 are explained in more detail below and Figure 5.2 translates this diagram into a set of template sub-processes which could be tailored for implementation within the organisation to achieve 'readiness for KLM'.

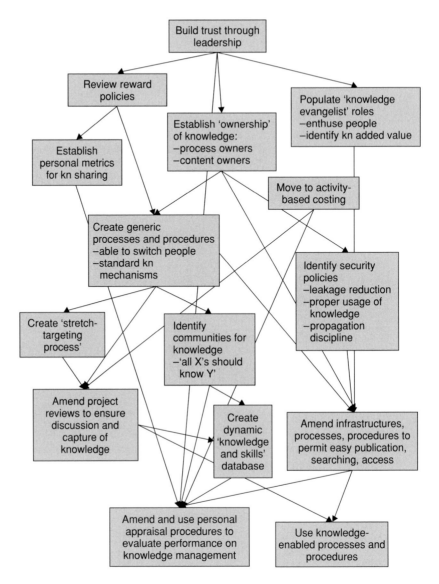

Figure 5.1   Key issues in readiness for knowledge lifecycle management (Kn, knowledge)

5.3.1   *Build trust through leadership*

This applies especially at Board level, but also throughout the firm's levels. There must be evident commitment to the philosophy of partnership in external relations and quality performance. Without this guidance there is a danger that process teams will overdevelop their own working culture to fill the gap, with the danger that this culture will address the dilemmas of the moment, rather than the longer-term strategic needs of the business. Also,

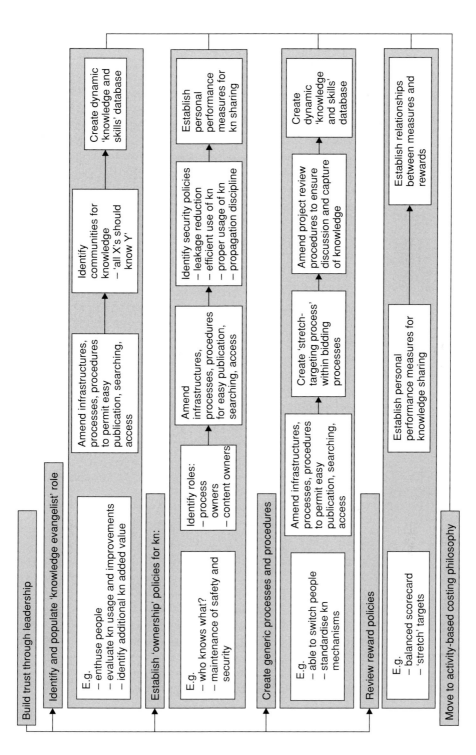

Figure 5.2   Readiness sub-processes (Kn, knowledge)

unless there is evident commitment from the Board, the initial impetus will come to a stop amidst much cynicism and complaint. Leadership cannot be delegated; leaders must be seen to lead.

5.3.2   Identify and populate 'knowledge evangelist' roles

This is a necessary step, because although we have just argued the need for leaders to be seen to lead, they cannot be everywhere, and a cadre of evangelists to propagate the leaders' message will be important to create committed disciples instead of acquiescent followers.

5.3.3   Establish 'ownership' policies for knowledge

The important roles here are those of process owners, content owners and process operators (Hammer, 1996; Siemieniuch and Sinclair, 2000). Ownership of processes has particular advantages:

- To document and 'proceduralise' the process, its controls and resource requirements in a generic form that represents best current practice.
- To ensure that this generic process is maintained as best current practice in its various instantiations in different facilities.
- To authorise variations from the generic process to fit local circumstances in a given instantiation.
- To accept and authorise improvements to the process as an ongoing evolutionary strategy.
- To ensure that any proposed changes to the process do not have deleterious effects on other, related processes (and vice versa).
- To support the change process by which process improvements are introduced.

These constitute ways in which important knowledge can be carried forward into the future; obviously, some co-ordination of processes across the e-supply configuration will be necessary, but this will have been considered above. Content owners provide knowledge outside the process; design knowledge about products, for instance. Process operators are the source from which new knowledge will be grown, and who embody the tacit knowledge by which formal knowledge is implemented in a given process in a given layout in a given organisational unit.

5.3.4   Identify and implement workable security policies

A critical organisational issue here is that security should not compromise the performance of people. If it does, it will not take long for security to be compromised. The authors have experience of a top-security site where the door codes were changed centrally every 6 hours. Because of the inconvenience caused by this, it took little time for it to become accepted practice for a set of current numbers to be taped near to each door lock, useful to insiders and visible to outsiders. It should be remembered that security systems are only as secure as the people who use them, and it is better to assume that the purpose of security is to slow the rate of knowledge leakage, not to keep secrets.

5.3.5 Create generic processes and procedures

These provide the basic framework of the company's operations, and through which it is managed and resourced. There are three significant points about generic processes; first, an efficient e-supply context is predicated on common ways of working and common standards; second, it enables work to be moved to the people wherever they are in the world; third, it enables different company units to present a common face to the same customers and suppliers, avoiding confusions and misunderstandings which can seriously undermine trust.

5.3.6 Amend technical infrastructures and processes to permit easy access, searching, publication and utilisation of knowledge

Some of this overlaps with the discussion above; it requires prior consideration of the aims of KLM (i.e. 'why are you doing it?'), the development of policies and procedures, and the identification of knowledge users and stakeholders. Then KLM processes must be developed: (1) processes to accrete, apply, propagate and retire knowledge; (2) communication paths, mechanisms and media; (3) storage mechanisms; (4) registry and search facilities; and (5) processing tools. Finally, the population and maintenance of this KLM structure must be addressed: (1) knowledge sources and (2) control and monitoring mechanisms. An excellent example of this has been provided by Vasey (2000); since this is not easily accessible, the main points are summarised below:

- All knowledge and information must be owned by individuals throughout the company, as a matter of policy. This provides a source for the context, hidden meanings and reliability of the material, and a basis for security and management of the status of the material.
- Distributed publishing is essential, but there must be a managed process for this in order to ensure quality input to knowledge repositories.
- Flexible information management is required that is appropriate to the local organisational culture, otherwise the process will be resisted.
- Teams will usually generate and use their own databases, but KLM requires more than this; the database should be usable and used by the

whole community in order to get proper value from the contents. For example, experts will not use it and contribute to it because they already know most of the contents of value; besides, if they do provide content, they will be inundated with requests. Therefore, there must be additional benefit; for example, include external information, perhaps by mining other companies' public websites.

- Current best practice indicates that not much time and other resources should be spent on classifications of knowledge. Instead, it is better to use a sophisticated search engine. A justification for this is that one cannot predict the future usage of knowledge and information and therefore it is unlikely that it will be stored effectively for the future user population; therefore, do not waste time on this.

- Searching must honour ownership and security. In other words, KLM must be an integral part of the IT infrastructure, and subject to 'good management practice'.

- Organisational leadership and policies are required to create a culture of 'give to get'. A discussion of some requirements for this can be found in Siemieniuch and Sinclair (1999).

- Create a 'who knows what' directory (a skills matrix, with personal editing to ensure it is current) as a first activity. In most business processes, a recurrent question is 'who knows about X?', and such a directory can provide an immediate, real, benefit to sceptical employees and managers at little cost, supplementing the usual, informal grapevine approach.

- Provide facilities *for* people, not *to* them. This is another way of emphasising the principle of 'user-centred design', and may be translated as do not ask users to do more work for KLM; instead, adapt normal working practices to permit KLM. For example, ensure that all content is saved on a server, rather than desk-top media, and this should be no harder to accomplish than current save procedures.

- Use text mining to summarise information and generate new relationships; furthermore, ensure there are good visual front-end tools to present the results to users. By mining internet news sites, for example, it has been shown to be possible to identify the marketing strategies of competitors; by mining intranet resources, it has been possible to find internal solutions to problems on current projects.

- Measures for the usage of intellectual capital are vital, and they must be dynamic ('how are we doing today?'). Examples of possible metrics can be found in Kaplan and Norton (1996) and Edvinsson and Malone (1998); Verkasalo and Lappalainen (1998) present a novel approach.

- The final point made by Vasey (2000) is that KLM does not remove the need for people, and it does not mean that you no longer need experts; experience of KLM demonstrates that it improves the richness of communications between people – they become more valuable, and knowledge is spread wider. In most cases, this increase in communications traffic is unlikely to bring about the stasis identified by complexity authors, because the levels of communication currently are so low. Where this becomes a problem, solutions will inevitably arise (e.g. 'read only those emails from known people, and junk the rest').

5.3.7 Review reward policies

Again, there is vast literature on this subject – compare, for example, Eason (1988); Ehn (1988); Huczynski and Buchanan (1991); Becker (1992); Limerick *et al.* (1994); Davenport *et al.* (1996); Kaplan and Norton (1996); Bradley (1997); Draycott (1997); Davenport and Prusak (1998); Edvinsson and Malone (1998). The authors of this chapter, as do several of the authors above, cling to the Occidental view that if the work that people do is not rewarded, it will be deferred, dropped or done badly. Hence, we argue that appropriate performance measurement criteria, reward structures and training provision are necessary to encourage people to work towards the goals of the company and in a trustworthy manner. These should be consistent (but not necessarily the same for everybody) within the organisation. Furthermore, the rewards do not necessarily have to be financial; Caterpillar's 'Wall of Patents' is an example of this, recognising in public the contributions of its personnel to knowledge growth.

5.3.8 Use personal appraisal procedures to evaluate performance on KM

The use of personal appraisal for performance assessment is now common; however, usually it takes little account of people's contributions to KLM. As the discussion has indicated, an individual's contribution to KLM can come in many forms, from indirect contributions such as leadership to direct contributions such as devising an improved process. Both of these classes of contributions should be assessed, as is discussed in particular by Kaplan and Norton (1996) and Skandia (1995, 1997, 1998).

5.3.9 Identify communities of knowledge

As Grudin (1994) demonstrates, there is little point in having superb IT infrastructures if there is no useful content. This comes back to people, who have to be willing, even eager, both to use the system and to contribute to its content. There are three aspects to this: first, people will not continue to contribute unless there is some reciprocal return and recognition of their efforts. Vasey's contributions (Vasey, 2000) are apposite here, as are peer appraisals. Second, there is a need for such systems to be user-friendly; not just that they have good IT interfaces, but that the knowledge is in a usable and useful form for its intended users. This is a difficult area, requiring further research. Third, the generation of peer group networks propagates the flow of expertise and other tacit knowledge (Snowden, 1998). By definition, these cannot easily be captured in IT infrastructures, and therefore require the creation and encouragement of such networks, transcending the boundaries of empowered teams, and going beyond the boundaries of individual companies and their value constellations (this is particularly true of networks of qualified professionals).

5.3.10   Move to an activity-based costing approach

How this should be accomplished is for each firm to decide. What is clear is that if companies do not have a clear understanding of the costs and benefits of their activities, they are unlikely to make good progress into the future competitive environment (Berliner and Brimson, 1988). Important within this is the measurement of human capital, something that still defies clear analysis, as is indicated in an excellent discussion by Abdolmohammadi *et al.* (1998).

5.3.11   Create a 'stretch-targeting process' within the contractual process

The purpose of this within an e-supply chain is two-fold: first, to ensure planned (i.e. 'no surprises') competitive progress by the value constellation in order to match or exceed likely developments by competitors; and second, to provide development targets within the individual companies that will maximise their knowledge growth and capability acquisition. Allied to benchmarking, this can be an effective way of introducing new thinking into an organisation.

5.3.12   Amend project review procedures to ensure discussion and capture of knowledge

This is a key part of KLM; however, review processes are planned, whether as 'stage-gate' phases within projects (Cooper, 1990; Cooper and Kleinschmidt, 1993, 1995) or as regular reviews of performance in operations.

5.3.13   Create dynamic knowledge and skills databases

These have been discussed by Vasey (2000), but it is stressed that they should supplement the usual human grapevines, rather than supplant them.

5.4   Conclusions

It is self-evident that knowledge lifecycle management is not merely a question of buying in sophisticated knowledge elicitation and mining technologies, although these tools do have a critical role to play in the overall KLM process. This chapter has also tried to make it clear that for an organisation to get the most added value out of the knowledge that is held within the organisation (in people's heads, in design databases, in processes and products, etc.), it must establish an appropriate context for KM.

Organisations rarely take the time to evaluate whether they are in fact in a position to implement policies and procedures to manage knowledge, and this chapter has set out some generic issues and possible steps that organisations need to address in this area. Since each organisation is an individual entity it is only possible to highlight generic issues and suggest template processes that need to be tailored to individual needs. It is intended that the process steps could be used as a kind of check-list to ascertain the current state of an organisation, but it is important to highlight that few if any organisations will be able to tick every box prior to moving down a KLM route. As stated in Section 5.2, many of the organisational infrastructural issues covered in this chapter are also relevant to the construction of all seven pillars on which organisational effectiveness depends. This is not surprising since KLM is certainly a fundamental support process to all other operational and management processes within organisations. Neither is this approach promulgated as 'right first time'. Preparing an organisation for KLM is an interactive learning process in its own right.

References

Abdolmohammadi, M.J., Greenlay, L. and Poole, D.V. (1998) *Accounting methods for measuring intellectual capital, round table group.* www.round.table.com/scholars/articles/acctg-intellectual-capital.html (accessed 12 Feb 2004).

Badham, R., Couchman, P. and Zanko, M. (2000) Implementing concurrent engineering. *Human Factors and Ergonomics in Manufacturing*, **10**(3), 237–50.

Becker, G.S. (1992) Nobel lecture: the economic way of looking at behaviour. The Nobel Foundation, Sweden. See *Journal of Political Economy*, June 1993, 385–409.

Berliner, C. and Brimson, J.A. (1988) *Cost Management for Today's Advanced Manufacturing*. Harvard Business School Press, Boston.

Bradley, K. (1997) Intellectual capital and the new wealth of nations, part II. *Business Strategy Review*, **8**(4), 33–44.

Cooper, R.G. (1990) Stage-gate systems: a new tool for managing new products. *Business Horizons*, May–June, 44–54.

Cooper, R.G. and Kleinschmidt, E.J. (1993) Screening new products for potential winners. *Journal of Long Range Planning*, **26**(6), 74–81.

Cooper, R.G. and Kleinschmidt, E.J. (1995) Benchmarking firms' new product performances and practices. *Engineering Management Review*, **23**(3), 112–20.

Craig, T. (1998) *Supply chain management: six issues that impact its effectiveness.* http://www.round.table.com/scholars/articles/supply-chain.html, Aug 1998 (accessed 12 Feb 2004).

Crosby, P.B. (1979) *Quality Is Free: The Art of Making Quality Certain.* McGraw Hill, New York.

Davenport, T.H. and Prusak, L. (1998) *Working Knowledge: How Organisations Manage What They Know.* Harvard Business School Press, Boston.

Davenport, T.H., Jarvenpaa, S.L. and Beers, M.C. (1996) Improving knowledge work processes. *Sloan Management Review,* Summer, 53–65.

Denton, J. (1998) *Organisational Learning and Effectiveness.* Routledge, London.

Despres, C. and Chauvel, D. (1999) Knowledge management(s). *Journal of Knowledge Management,* **3**(2), 110–20.

Draycott, P. (1997) People: aligning strategy with people. *Maintenance and Asset Management,* **12**(5), 18–25.

Eason, K.D. (1988) *Information Technology and Organisational Change.* Taylor & Francis, London.

Edvinsson, L. and Malone, M.S. (1998) *Intellectual Capital: The Proven Way to Establish Your Company's Real Value by Measuring its Hidden Brainpower.* Piatkus, London.

Ehn, P. (1988) *The Work Oriented Design of Computer Artifacts.* Arbetsmiljo, Arbelistraum, Centre for Working Life, Stockholm.

Gregg, D. (1996) *Emerging Challenges in Business and Manufacturing Decision Support. The Science of Business Process Analysis.* ESRC Business Process Resource Centre, University of Warwick, Coventry.

Grudin, J. (1994) Computer-supported co-operative work: history and focus. *Computer* **27**(5), 19–26.

Hammer, M. (1996) *Beyond Re-engineering.* Harper Collins, New York.

Hampden-Turner, C. and Trompenaars, A. (1994) *The Seven Cultures of Capitalism.* Piatkus, London.

Handy, C.B. (1976) *Understanding Organisations.* Penguin, London.

Handy, C. (1989) *The Age of Unreason.* Hutchinson, London.

Handy, C. (1994) *The Empty Raincoat.* Hutchinson, London.

Haywood, B. and Barrar, P. (1998) *Manufacturing Company Supply Chains: SMEs in the North West.* Manchester Business School, Manchester.

Helmreich, R.L. and Merritt, A.C. (1998) *Culture at Work in Aviation and Medicine.* Ashgate, Aldershot.

Hofstede, G. (1984) *Culture's Consequences: International Differences in Work-Related Values.* Sage, Beverley Hills.

Hofstede, G. (1991) *Cultures and Organisations.* McGraw-Hill, London.

Huczynski, A. and Buchanan, D. (1991) *Organisational Behaviour.* Prentice Hall, London.

Kamara, J.M., Anumba, C.J., *et al.* (2001) Selection of a knowledge management strategy for organisations. In: *Proceedings of European Conference on Knowledge Management (ECKM-2001)* (D. Remenyi, ed.), Bled, Slovenia, pp. 243–54.

Kaplan, R.S. and Norton, D.P. (1996) *The Balanced Scorecard.* Harvard Business School Press, Boston.

Kotter, J.P. and Heskett, J.L. (1992) *Corporate Culture and Performance.* Free Press, New York.

Leonard-Barton, D. (1995) *Wellsprings of Knowledge: Building and Sustaining the Sources of Innovation.* Harvard Business School Press, Boston.

Leseure, M. and Brookes, N.J. (2000) A support tool for knowledge management activities. *IEEE International Conference on Management of Innovation and Technology*, Singapore, pp. 696–701.

Levinson, N.S. and Asahi, M. (1995) Cross-national alliances and interorganisational learning. *Organisational Dynamics*, **24**(2), 50–64.

Limerick, D., Passfield, R. and Cunnington, B. (1994) Transformational change – towards an action learning organisation. *The Learning Organisation*, **1**(2), 29–40.

Lund, R.T., Bishop, A.B., Newman, E.E. and Salzman, H. (1993) *Designed to Work: Production Systems and People*. Prentice Hall, Englewood Cliffs, New Jersey.

Markus, M.L. (2001) Towards a theory of knowledge reuse: types of knowledge reuse situations and factors in reuse success. *Journal of Management Information Systems*, **18**(1), 57–93.

Meier, A. (1999) Occupational cultures as a challenge to technological innovation. *IEEE Transactions on Engineering Management*, **46**(1), 101–114.

Ministry of Defence (MOD, 1999) *The Acquisition Handbook: A Guide to Smart Procurement*. Ministry of Defence, Smart Procurement Implementation Team. HMSO, London.

Nadler, D.A. and Tushman, M.L. (1999) The organisation of the future: strategic imperatives and core competencies for the 21st century. *Organisational Dynamics*, **28**(1), 45–6.

OSTEMS (1998) *Understanding Best Practice for Key Elements of the New Product Introduction Process*. DTI/ESRC Business Processes Resource Centre, University of Warwick, Coventry.

Revans, R.W. (1982) *The Origins and Growth of Action Learning*. Bromley, T. Chartwell Bratt, London.

Rycroft, R.W. and Kash, D.E. (1999) *The Complexity Challenge*. Pinter, London.

Schein, E. (1996) Culture: the missing concept in organisation studies. *Administrative Sciences Quarterly*, **41**, 229–40.

Siemieniuch, C.E. and Sinclair, M.A. (1993) Implications of concurrent engineering for organisational knowledge and structure – a European, ergonomics perspective. *Journal of Design and Manufacturing*, **3**(3), 189–200.

Siemieniuch, C.E. and Sinclair, M.A. (1999) Organisational aspects of knowledge lifecycle management in manufacturing. *International Journal of Human–Computer Studies*, **51**, 517–47.

Siemieniuch, C.E. and Sinclair, M.A. (2000) Implications of the supply chain for role definitions in concurrent engineering. *International Journal of Human Factors and Ergonomics in Manufacturing*, **10**(3), 251–72.

Siemieniuch, C.E. and Sinclair, M.A. (2002) On complexity, process ownership and organisational learning in manufacturing organisations, from an ergonomics perspective. *Applied Ergonomics*, **33**(5), 449–62.

Siemieniuch, C.E., Waddell, F.N. and Sinclair, M.A. (1999) The role of 'partnership' in supply chain management for fast-moving consumer goods: a case study. *International Journal of Logistics: Research and Applications*, **2**(1), 87–101.

Skandia (1995) *Value - Creating Process*. Skandia AFS, http://www.skandia.se.

Skandia (1997) *Intelligent Enterprising*. Skandia AFS, http://www.skandia.se.

Skandia (1998) *Human Capital in Transformation*. Skandia AFS, http://www.skandia.se.

Snowden, D. (1998) A framework for creating a sustainable programme. In: *Knowledge Management: A Real Business Guide* (S. Rock and J. Reeves, eds), pp. 6–19. Caspian, London.

Stewart, T. A. (1997) *Intellectual Capital: The New Wealth of Companies*. Doubleday, New York.

Towill, D.R. (1996) Time compression and supply chain management – a guided tour. *Supply Chain Management*, **1**(1), 15–27.

Trompenaars, F. (1994) *Riding the Waves of Culture: Understanding Diversity in Global Business*. Irwin, New York.

Vasey, M. (2000) Strategic value of knowledge in an environment of change. *Presentation at EPSRC/SEBPC/British Academy of Management Workshop: Knowledge Management for Strategic Business Change*, Gas Research and Technology Centre, BG Technology, Loughborough.

Verkasalo, M. and Lappalainen, P. (1998) A method of measuring the efficiency of the knowledge utilization process. *IEEE Transactions on Engineering Management*, **45**(4), 414–23.

6 Tools and Techniques for Knowledge Management

Ahmed M. Al-Ghassani, Chimay J. Anumba,
Patricia M. Carrillo and Herbert S. Robinson

6.1 Introduction

The term 'knowledge management tools' is sometimes used narrowly to mean information technology (IT) tools. Knowledge management (KM) tools are both IT and non-IT tools required to support sub-processes of KM such as locating, sharing and modifying knowledge. There is a large number of tools available to choose from in implementing a KM strategy. However, organisations often encounter difficulties in identifying appropriate tools due to the range of competing products in the marketplace, claims of vendors, overlap between the functions of various tools and high cost often associated with acquiring and using them. Selecting appropriate tools is, therefore, vital for implementing a KM strategy, but careful investigation is required to ensure that the business problems and context are understood and the organisational goals are adequately addressed.

This chapter focuses on the role of KM tools in the implementation process. Following the Introduction, Section 6.2 presents a comparison and a description of various KM tools, distinguishing clearly between KM techniques and technologies. Section 6.3 examines existing approaches for selecting tools to support KM and discusses the limitations of the methods. An innovative method (SeLEKT) is presented in Section 6.4, as an alternative, more structured and informed approach. The method facilitates the selection of appropriate tools based on key dimensions of KM that reflect an organisation's business needs and context. The key dimensions are: the 'knowledge transfer domains', that is, whether the knowledge is internal or external to the organisation; 'ownership forms', reflecting whether knowledge is owned by individuals or shared; and the 'conversion types', reflecting the interactions between tacit and explicit knowledge. The chapter concludes (Section 6.5) that the SeLEKT framework is a useful and practical approach for selecting the most appropriate tools and applications to support KM based on an organisation's context and business requirements.

Table 6.1   A comparison between KM techniques and technologies

KM tools	
KM techniques	**KM technologies**

• Require strategies for learning
• More involvement of people
• Affordable to most organisations
• Easy to implement and maintain
• More focus on tacit knowledge
• Examples of tools:
   ○ Brainstorming
   ○ Communities of practice
   ○ Face-to-face interactions
   ○ Recruitment
   ○ Training
	• Require IT infrastructure
	• Require IT skills
	• Expensive to acquire/maintain
	• Sophisticated implementation/maintenance
	• More focus on explicit knowledge
	• Examples of tools:
	   ○ Data and text mining
	   ○ Groupware
	   ○ Intranets/extranets
	   ○ Knowledge bases
	   ○ Taxonomies/ontologies

6.2   Knowledge management tools

Very few authors have defined KM tools. Gallupe (2001) states that they are not simply information management tools as they should be 'capable of handling the richness, the content, and the context of the information and not just the information itself'. A popular definition by Ruggles (1997) describes them as the technologies used to enhance and enable the implementation of the sub-processes of KM, e.g. knowledge generation, codification and transfer. He argued that not all KM tools are IT based, as a paper, pen or video can also be utilised to support KM. To distinguish between KM tools, the terms 'KM techniques' and 'KM technologies' are used to represent 'non-IT tools' and 'IT tools' respectively. The main differences between KM techniques and technologies are presented in Table 6.1 and discussed thereafter.

6.2.1   Techniques

KM techniques (non-IT tools) are tools that do not require technology to support them and exist in several forms. For example, knowledge sharing, a sub-process of KM, can take place through face-to-face meetings, recruitment, seminars, etc. KM techniques are important due to several factors. First, they are affordable to most organisations, as no sophisticated infrastructure is required (although some techniques require more resources than others, e.g. training requires more resources than face-to-face interactions). Second, KM techniques are easy to implement and maintain as they incorporate features that are relatively simple and straightforward to understand. Finally, they focus on retaining and increasing the organisational tacit knowledge, which is a key asset to organisations.

KM techniques are not new, as organisations have been implementing them for a long time, but their use has been under the umbrella of several management approaches. Using these tools for the management of organisational knowledge requires their use to be enhanced so that benefits from them, in terms of knowledge gain/increase, are properly managed. Examples of some KM techniques are described below:

Brainstorming is a process involving a group of people who meet to focus on a problem, and then intentionally propose as many deliberately unusual solutions as possible through pushing the ideas as far as possible. The participants shout out ideas as they occur to them and then build on the ideas raised by others. All the ideas are noted down and are not criticised. Only when the brainstorming session is over are the ideas evaluated. Brainstorming helps in problem solving and in creating new knowledge from existing knowledge (Tsui, 2002a,b).

Communities of practice (CoPs) are also called knowledge communities, knowledge networks, learning communities, communities of interest and thematic groups. These consist of a group of people of different skill sets, development histories and experience backgrounds that work together to achieve commonly shared goals (Ruggles, 1997). These groups are different from teams and task forces. People in CoPs can perform the same job, collaborate on a shared task (software developers) or work together on a product (engineers, marketers and manufacturing specialists). They are peers in the execution of 'real work'. What holds them together is a common sense of purpose and a real need to know what each other knows. Usually, there are many CoPs within a single company and most people normally belong to more than one.

Face-to-face interaction is a traditional approach for sharing the tacit knowledge (socialisation) owned by an organisation's employees. It usually takes an informal approach and is very powerful. Face-to-face interaction also helps in increasing the organisation's memory, developing trust and encouraging effective learning. Lang (2001) considers it to provide strong social ties and tacit shared understandings that give rise to collective sense-making. This can also lead to an emergent consensus as to what is valid knowledge and to the serendipitous creation of new knowledge and, therefore, new value. This provides an environment within an organisation where participants see the firm as a human community capable of providing diverse meaning to information (i.e. knowledge).

Post-project reviews are debriefing sessions used to highlight lessons learnt during the course of a project. These reviews are important to capture knowledge about causes of failures, how they were addressed, and the best practices identified in a project. This increases the effectiveness of learning as knowledge can be transferred to subsequent projects. However, if this technique is to be effectively utilised, adequate time should be allocated for those who were involved in a project to participate. It is also crucial for post-project review meetings to take place immediately after a project is completed as project participants may move or be transferred to other projects or organisations.

Recruitment is an easy way to buy-in knowledge. This is a technique for acquiring external tacit knowledge especially of experts. This approach adds new knowledge and expands the organisational knowledge base. Another benefit is that other members within the organisation can learn from the recruited member both formally and informally so that some knowledge will be transferred and retained if the individual leaves the organisation. Some organisations also try to codify the recruited person's knowledge that is of critical importance to their business.

Apprenticeship is a form of training in a particular trade carried out mainly by practical experience or learning by doing (not through formal instruction). Apprentices often work with their masters and learn craftsmanship through observation, imitation and practice. The masters focus on improving the skills of the individuals so that they can later perform tasks on their own. This process of skill building requires continuous practice by the apprentices until they reach the required level.

Mentoring is a process where a trainee or a junior member of staff is attached or assigned to a senior member of an organisation for advice related to career development. The mentor provides a coaching role to facilitate the development of the trainee by identifying training needs and other development aspirations. This type of training usually consists of career objectives given to the trainee and the mentor checks if the objectives are achieved and provides feedback.

Training helps to improve staff skills and therefore increase knowledge. Its implementation depends on plans and strategies developed by the organisation to ensure that employees' knowledge is continuously updated. Training usually takes a formal format and can be internal, where senior staff train junior employees within the organisation, or external, where employees attend courses managed by professional organisations.

6.2.2 Technologies

KM technologies depend heavily on IT as the main platform for implementation. Examples of KM technologies for capturing knowledge are knowledge mapping tools, knowledge bases and case-based reasoning. Although there is a debate about the degree of importance of such technologies, many organisations consider them as important enablers to support the implementation of a KM strategy (Anumba *et al.*, 2000; Egbu, 2000; Storey and Barnet, 2000). KM technologies consume about one third of the time, effort and money that are required for a KM system and the other two thirds relate mainly to people and organisational culture (Davenport and Prusak, 1998; Tiwana, 2000). Ruggles (1997) relates the importance of IT tools to their quick evolution, dynamic capabilities and high cost. KM technologies consist of a combination of hardware and software technologies.

Hardware technologies are very important for a KM system as they form the platform for the software technologies to perform and the medium for the storage and transfer of knowledge. Some of the hardware requirements for a KM system are (Lucca *et al.*, 2000):

- personal computer or workstation to facilitate access to the required knowledge
- highly powerful servers to allow the organisation to be networked
- open architecture to ensure interoperability in distributed environments
- media-rich applications requiring integrated services digital network (ISDN) and fibre optics to provide high speed
- asynchronous transfer mode (ATM) as a multimedia switching technology for handling the combination of voice, video and data traffic simultaneously
- use of the public network technology (e.g. internet) and private network technologies (e.g. intranet, extranet) to facilitate access and sharing of knowledge.

Software technologies play an important part in facilitating the implementation of KM. The number of software applications has increased considerably in recent years. Solutions provided by software vendors take many forms and can perform different tasks. The large number of vendors that provide KM solutions makes it extremely difficult to identify the most appropriate applications. This has resulted in organisations adopting different models for establishing KM systems. Tsui (2002b) identifies five emerging models for deploying organisational KM systems where one or a combination may be adopted:

(1) *Customised off-the-shelf (COTS)*. This is the traditional and most popular way of deploying application services. Based on the organisational needs, the applications will be identified and then examined against the functional needs of the organisation. A short test period may follow to identify the most suitable application. Once an application is acquired, customisations on the standard features are usually performed to integrate it into the organisation's system.

(2) *In-house development*. These systems are developed within the organisation, usually with external technical help. Examples are Notes, Domino and intranet applications. However, several reasons make this option generally less attractive or preferred by organisations. This includes the difficulty of establishing KM system requirements, high cost, risk and the complexity often associated with developing bespoke systems.

(3) *Solution re-engineering*. This involves adapting, with the help of KM consultants and technical architects, an existing generic solution that matches the organisation's requirements. Although similar to COTS,

the adapted solution is not packaged as a product that can be marketed. Examples are online knowledge communities and virtual collaboration tools.

(4) *Knowledge services*. These are knowledge applications provided by a third party that hosts the application on the web. The user accesses the service via a thin-client (e.g. a browser). The main benefits are the waived software licensing fee and the avoidance of in-house maintenance. However, many organisations do not find this option attractive because of the reduced security and privacy.

(5) *Knowledge marketplace*. Modelled in the e-business NetMarket concept, several knowledge-trading places have been established recently. In a knowledge marketplace, a third-party vendor hosts a website grouping together many suppliers of knowledge services. Suppliers may include expert advisors, vendors providing product support services, KM job placement agencies, procedures of evaluations of KM and portal software, and research companies providing industry benchmarks and best practice case studies. Two types of knowledge marketplace exist: one provides common information and services to all industries while the other offers only certain services to a specific industry.

KM software technologies have undergone significant improvements in the last few years as a result of alliances, and mergers and acquisitions between KM and portal tool vendors (Tsui, 2002b). However, no software technology can provide a complete solution to KM. A number of tools are better described as KM technologies such as data and text mining, groupware, etc. Some of these are described below.

Data and text mining is a technology for extracting meaningful knowledge from masses of data or text. Data are single facts (structured) about events while text refers to unstructured data. The process of data/text mining enables meaningful patterns and associations of data (words and phrases) to be identified from one or more large databases/knowledge bases. The approach is very useful for identifying hidden relationships between data and hence creating new knowledge. It is mostly used in business intelligence, direct marketing and customer relationship management applications. However, its application is limited due to the difficulties in accessing data via an enterprise-wide corporate portal where most organisations only have a small group of data miners (Tsui, 2002a,b).

Groupware is a software product that helps people to communicate, share information, perform work efficiently and effectively, and to facilitate group decision-making using IT (Haag and Keen, 1996). It supports distributed and virtual project teams where team members are from multiple organisations and in geographically dispersed locations. Groupware tools usually contain email communication, instant messaging, discussion areas, file area or document repository, information management tools (e.g. calendar, contact lists, meeting agendas and minutes) and search facilities.

Intranet is an internal organisational internet that is guarded against outside access by special security tools called firewalls, while *extranet* is an intranet with limited access to outsiders, making it possible for them to collect and deliver certain knowledge on the intranet. This technology is very useful for making organisational knowledge available to geographically dispersed staff members and is therefore used by many organisations.

Knowledge bases are repositories that store knowledge about a topic in a concise and organised manner. They present facts that can be found in a book, a collection of books, websites or even human knowledge. This is different from the part of the expert system/case base reasoning (CBR) that stores the rules.

Taxonomy is a collection of terms (and the relationships between them) that are commonly used in an organisation. Examples of a relationship are 'hierarchical' (where one term is more general and hence subsumes another term), 'functional' (where terms are indexed based on their functional capabilities) and 'networked' (where there are multiple links between the terms defined in the taxonomy). *Ontologies* define the terms and their relationships, but, in addition, they support deep (refined) representation (for both descriptive and procedural knowledge) of each of the terms (concepts) as well as defined domain theory or theories that govern the permissible operations with the concepts in the ontology. There are at least three ways to develop a taxonomy/ontology: manually constructed (using some kind of building tools), automatically discovered (from a repository of knowledge assets) or purchased off the shelf. Taxonomies/ontologies serve multiple purposes in an organisation. They can be used as a corporate glossary, holding detailed descriptions of every key term used in the organisation. They can also be used to constrain the search space of search engines and refine search results, identify and group people with common interests, and act as a content/knowledge map to improve the compilation and real-time navigation of web pages (Tsui, 2002a,b).

6.3   Selecting knowledge management tools

The technologies described above are just examples of some of the more common KM tools. The software technology market is very dynamic and is continually evolving with better and more refined products to support KM. This situation creates problems for selecting the most appropriate tools but also presents opportunities for organisations to be able to access state-of-the-art technology to manage their most valuable knowledge resource. A recent survey of construction organisations carried out at Loughborough University shows that the most widely used technology is the intranet (Carrillo *et al.*, 2004). Intranets provide the platform for knowledge sharing particularly in large construction organisations that are often geographically dispersed with diverse knowledge to share.

Other popular technologies are document management systems (e.g. Documentum and Sage Desk), groupware (e.g. Lotus Notes, Lotus Quickplace, Live Link and e-Room) and taxonomy tools (e.g. Autonomy). The use of technologies may increase as collaborative working becomes more important in the construction supply chain. Extranets and electronic discussion forums are used to a limited extent. However, it is expected that the use of the extranets will increase, as new procurement approaches involving the entire construction supply chain such as partnering become more important.

Communities of practice is the most widely used technique for KM particularly in large organisations. Large construction organisations with a range of specialist skills tend to have the need and resources to set up communities of practice and to benefit significantly from them. Other techniques that are used include brainstorming, job observation and rotation systems, research collaboration, conferences and seminars to facilitate knowledge sharing and to update knowledge.

These techniques and technologies provide valuable support for KM. However, it is often difficult to select the most appropriate tools for a particular organisation. The large number of technologies available in the market place makes it difficult for organisations to identify the most appropriate tools, and construction organisations are no exception. The next section presents and discusses classification of KM tools, which reflects existing approaches for selecting the most suitable tools. The selection of KM tools requires a clear identification of the organisational needs from KM and at the same time requires knowing the different tools available and their functional capabilities. Existing methods for selecting KM tools vary from one organisation to another.

6.3.1 Existing methods

When developing a KM strategy, organisations are required to identify the techniques (non-IT tools) and technologies (IT tools) needed. This identification is associated with difficulties relating to the location and nature of knowledge, organisational goal(s) for implementing KM, costs and the overlap between what these tools can do.

Selecting KM techniques

Selecting the most appropriate KM techniques does not usually follow a structured approach. This is usually due to company traditional practices, the ease of implementation and the relatively low cost associated with setting up and maintaining KM techniques, compared to technologies. Also, most organisations using particular forms of KM techniques may have improved their implementation over time. This often leads to the preference of these techniques without a thorough investigation of the alternative options or their input to the KM system. The result is often the adoption of techniques that may not be too critical for the system, at the expense of exploring other more applicable ones.

Selecting KM technologies

The selection of KM technologies is far more complicated. It requires a clear identification of the organisation's KM needs and an awareness of the technologies available, their cost and functional capabilities. For these reasons, most researchers and practitioners tend to focus on the selection of appropriate KM technologies. There are two main approaches to selection: according to KM sub-processes and according to technology families. The former categorises the tools in terms of the KM sub-processes they support while the latter classifies them into general technology families that support KM.

Selection of KM technologies according to KM sub-processes
This method is popular because it allows users to identify the sub-processes that they need to manage and then select the most appropriate technologies for them. After identifying the KM sub-processes, authors split into two groups. The first group (Table 6.2) identified the commercial KM software applications that support the KM sub-processes without putting them into technology categories. Ruggles (1997) was one of the first to follow this pattern. Wensley and Verwijk-O'Sullivan (2000) adopted the same pattern but considered 'web-based' software applications and produced a long list containing products developed by 164 software vendors.

Table 6.2   Software applications classified by KM sub-processes

Author	KM sub-processes	KM software applications
Ruggles (1997)	● Generation	GrapeVine, IdeaFisher, Inspiration, Idea Generator, MindLink
	● Codification	KnowledgeX, Excalibur RetrievalWare and Visual RetrievalWare, TeleSim
	● Transfer	(Lotus) Notes, NetMeeting, EnCompass
Wensley and Verwijk-O'Sullivan (2000) (web-based KM tools)	● Acquire	Aeneid, Networker, Infoscout, Arbortext tools, Documentum
	● Store	2 Share 2.0, Beehive, Action Technologies Tools, WebOS, Aeneid, Networker, Infoscout, Arbortext tools, Autonomy, Documentum
	● Deploy	2 Share 2.0, Beehive, Action Technologies Tools, WebOS, Networker, Infoscout, Arbortext tools, Autonomy, Documentum
	● Add value	Action Technologies Tools, WebOS, Autonomy, Documentum

Table 6.3   KM technologies classified by KM sub-processes

Author	KM sub-processes	KM technologies
Jackson (1998)	• Gathering	Pull, searching, data entry/optical character recognition (OCR)
	• Storage	Linking, indexing, filtering
	• Communication	Sharing, collaboration, group decisions
	• Dissemination	Push, publishing, notification
	• Synthesis	Analysis, creation, contextualisation
Laudon and Laudon (2000)	• Creation	Knowledge work systems: computer-aided design (CAD), virtual reality, investment workstations
	• Knowledge capturing and codifying	Artificial intelligence systems: expert systems, neural nets, fuzzy logic, genetic algorithms, intelligent agents
	• Knowledge distribution	Office automation systems: word processing, desktop publishing, imaging and web publishing, electronic calendars, desktop databases
	• Knowledge sharing	Group collaboration systems: groupware, intranets
Tsui (2002a) (PKM tools)	• Creation	Associative links, information capturing and sharing, concept/mind mapping
	• Codification/ representation	Associative links, information capturing and sharing, concept/mind mapping, e-mail management, analysis and unified
	• Classification/indexing	Index/search, meta-search, associative links, information capturing and sharing, concept/mind mapping, e-mail management, analysis and unified
	• Search and filter	Index/search, meta-search, e-mail management, analysis and unified
	• Share/distribute	Index/search, meta-search, associative links, information capturing and sharing, e-mail management, analysis and unified

The second group (Table 6.3) identified categories of KM technologies that support the KM sub-processes without naming the software applications. They also included some technology categories that were not originally developed for KM but support its sub-processes. Jackson (1998) and Laudon and Laudon (2000) adopted this pattern. A similar attempt by Tsui (2002a) focuses on the personal KM technologies (PKM) rather than the enterprise KM technologies. Tsui adapted his approach from Barth (2000) and extended it to take account of changes in the market, e.g. corporate mergers, fall of dotcoms, etc.

Selection of KM technologies according to KM technology families
Technology families are categories of commercial KM software applications such as document management, groupware, search facilities, etc. Table 6.4 shows examples of classification by technology families.

Jackson's (1998) classification presents six technology families and identifies a few examples of commercial software applications for every category.

Table 6.4   KM technology families by different authors

Author	KM technology families
Jackson (1998)	Document management, e.g. Documentum, Panagon JetForm Information management, e.g. SAP, Baan Searching and indexing, e.g. Fulcrum, Retrievalware, Verity Communications and collaborations, e.g. Notes, Exchange, Eudora Expert systems, e.g. Trajecta, Cognos Systems for managing intellectual property
Wensley and Verwijk-O'Sullivan (2000) (web-based tools)	Traditional database tools Process modelling and management tools Workflow management tools Enterprise resource management tools Agent tools Search engines, navigation tools and portals Visualising tools Collaborative tools Virtual reality tools
Gallupe (2001)	Intranets Information retrieval programs Database management systems Document management systems Groupware Intelligent agents Knowledge-based or expert systems
Tsui (2002a,b)	Search Meta/web crawler Process modelling and mind mapping Case-based reasoning (CBR) Data and text mining Taxonomy/ontological tools Groupware Measurement and reporting E-learning

Wensley and Verwijk-O'Sullivan's (2000) classification focuses on web-based technologies. Gallupe's (2001) classification is based on a three-level model of KM systems. The three levels are tools, generators and the specific KM system. He identifies tools and generators as the technologies that are used to acquire, store and distribute knowledge. In this context, tools are basic technological building blocks for the KM system where individual tools can be combined to form a specific KM system that performs particular functions. On the other hand, generators are self-contained technologies and can be used to generate or build a variety of specific KM systems. A generator therefore consists of a number of tools such as document management, intelligent agent and groupware. For example, Lotus Notes is a generator that contains a number of KM features that can be combined in various ways to make different KM systems. Tsui's (2002b) classification is

Table 6.5   KM technology families by Bair and O'Connor (1998)

Information retrieval/ knowledge retrieval (IR/KR)	Document management (DM) Groupware (GW)	Integrated systems KR+DM+GW+data management
Fulcrum		IBM – Lotus
Dataware	IBM – Lotus	(1999 onwards)
Verity	(1998 and prior)	
Excalibur		
OpenText – IDI	Documentum	
Lycos +	PCDOCS	
InMagic		
	FileNet	
CompassWare		
Sovereign Hill		
		Novell
Inference		
	Autonomy	
InXight	Intraspect	
Plumtree		
Perspecta	Firefly	Netscape
SageWare	Wiseware	
Wincite	GrapeVine	
Magnifi		
Individual	Group	Organisation

Strength → (left axis)

Capacity for collaboration over time and across the organisation

based on the origin of technologies, alignment with business processes and capabilities of the commercial KM software.

Bair and O'Connor (1998) followed this method but in a more detailed way (Table 6.5). They introduced three technology families and then categorised KM software applications into them through identifying software vendors and classifying them according to the capacity for collaboration over time and across the organisation.

6.3.2   Limitations of existing methods

Existing methods for identifying the most appropriate KM tools show that these approaches do not fulfil some of the requirements for developing a KM strategy. Limitations in the current methods are described below.

KM techniques

The most important limitations in existing methods for selecting KM techniques are:

● They do not follow a structured approach and are therefore open to many interpretations.

- They depend, in many cases, on improving existing techniques without investigating whether they are needed for the KM system.
- The way these techniques are selected does not necessarily link the selection process to the organisational goals for implementing KM.

KM technologies

The most important limitations in existing methods for selecting KM technologies are:

- Classifications according to KM sub-processes do not link the technology families to their commercial software applications. This is probably because KM tools are still in the infancy stage and hence any list of commercial software may become outdated in a very short time.
- Classifications according to technology families are generic and therefore not very useful for organisations that seek practical methods for identifying the most appropriate tools.
- Classifications by technology families also identify technology families without naming the software applications that support them, although, in some cases, reference to examples of applications is given. This is probably due to two reasons: the large and increasing number of software products and the overlap between their functions.
- The existing methods are easy to use but do not link the selection to the organisational requirements for KM.

The limitations identified in existing methods for selecting the most appropriate KM tools indicate a need for a new approach that addresses these limitations.

6.4 The SeLEKT approach

Research on KM tools shows that organisations encounter difficulties in identifying the most appropriate tools due to the large number of products in the marketplace and the overlap between what they can do. The limitations identified in the existing methods for selecting KM tools have led to the development of an integrated approach (SeLEKT) that relies on specific selection criteria and integrates the existing methods. The process of selecting the most appropriate KM tools is influenced by several factors.

The SeLEKT (Searching and Locating Effective Knowledge Tools) approach was developed within a three-year framework project entitled 'Knowledge Management for Improved Business Performance (KnowBiz)' at Loughborough University, UK. The approach was developed with the support of industrial collaborators, mainly from the construction industry. The methodology adopted in the development of this method involved: a continuous review of the literature to maintain awareness of current developments in the field; an adoption of an organisation-centred approach to identify the critical criteria (KM dimensions) required for the tool selection; and an interactive user-based evaluation of the framework.

SeLEKT is a three-step process. First, it is required to identify the KM dimensions that reflect the organisation's status in terms of current and required dimensions. Second, the user needs to determine the KM sub-processes required for the identified knowledge dimensions and to link them to tool categories. The third stage is for the IT tools only where the user is required to select the software applications that support the identified tool category. These stages are described in more detail below.

6.4.1   Identify organisational KM dimensions (stage 1)

This stage consists of three 'KM dimensions' required to identify the most appropriate tools. Three dimensions have been identified as critical for the selection of KM tools (Al-Ghassani *et al.*, 2002). These dimensions are the 'knowledge transfer domains' (internal–external), the 'knowledge ownership form' (individual–group) and the 'knowledge conversion types' (tacit–explicit). Every dimension should be investigated in terms of current and required status of the knowledge of interest.

Knowledge transfer domains (the 'internal–external' dimension)

The knowledge transfer domains investigate whether the knowledge is internal to the organisation (e.g. owned by its experts or exists in its knowledge bases) or external to it (e.g. with consultants). Knowledge exists either within an organisation or outside it. Internal knowledge may be easier and less expensive to manage whereas external knowledge is more difficult and expensive to obtain. The location of the existing knowledge and its target destination have an impact on the technique or technology to be used for managing this knowledge. For example, communities of practice and intranets are used to transfer knowledge internally within organisations whilst extranets are used to transfer knowledge between organisations, i.e. from internal to external or external to internal. Recruitment and the purchase of knowledge bases are techniques for transferring external knowledge to internal. The internet, on the other hand, facilitates the transfer of knowledge from external to internal domains, although this requires validating the knowledge of interest.

Knowledge ownership forms (the 'personal–shared' dimension)

This dimension investigates the interaction between individual (personal) and group (shared) knowledge. The techniques and technologies required for personal knowledge are different from those required for treating shared knowledge. For example, the e-mail technology helps in transferring knowledge from an individual to another individual or to a group of people, whilst mentoring is a technique for transferring knowledge and experience from an individual to another individual. In order to know which technique or technology is the most suitable for a particular organisation or for a particular case it is important to first know who has this knowledge and who requires it. Other issues such as security are usually associated with this dimension especially when it is required to transfer a critical personal knowledge to a group of people or to a public or semi-public domain.

Knowledge conversion types (the 'tacit–explicit' dimension)

The knowledge conversion types refer to the interaction between tacit (e.g. in people's heads or organisational processes) and explicit (e.g. codified in documents, drawings) knowledge identified by Nonaka (1995).

The way this conversion takes place depends on the current and required status of knowledge. Knowledge conversion types are important because KM techniques and technologies treat these types differently. For example, data mining technologies are suitable for searching within explicit knowledge, whilst expert system technologies are used for converting tacit knowledge to explicit, unlike face-to-face interaction which is a technique for converting one tacit knowledge to another tacit knowledge. Therefore, the selection of the most suitable KM technique/technology depends on the current and required type of knowledge. Table 6.6 represents the logical

Table 6.6 KM dimensions and their possible combinations

KM dimensions				Required dimensions							
Transfer domains				Internal				External			
		Ownership forms		Individual		Group		Individual		Group	
			Conversion types	Tacit	Explicit	Tacit	Explicit	Tacit	Explicit	Tacit	Explicit
Internal	Individual		Tacit								
			Explicit								
	Group		Tacit								
			Explicit								
External	Individual		Tacit								
			Explicit								
	Group		Tacit								
			Explicit								

Current dimensions (left vertical label)

combinations of these dimensions. The shaded area on the right bottom corner indicates that transferring knowledge from external to external is not a logical combination in this context.

6.4.2   Identify KM sub-processes required and link them to tool categories (stage 2)

KM is a process consisting of several sub-processes where the sub-processes do not necessarily follow a linear relationship. The KM sub-processes used in SeLEKT are: locating and accessing, capturing, representing, sharing, and creating new knowledge. The definitions of the sub-processes are given below.

After identifying the current and required KM dimensions, it is important to know what KM sub-process(es) are involved. For example, if knowledge is to be transferred from 'internal–individual–tacit' to 'internal–individual–tacit' as for the first combination cell in Table 6.6, then two KM sub-processes will be involved, namely 'locating and accessing' and 'sharing'. This means that we need to 'locate' the individual who has the knowledge of interest and to facilitate the 'sharing' of this knowledge to the other individual who requires it. 'Locating' can be through the organisation's experts directory or skills yellow pages – whether manual or automated – while 'sharing' may be through face-to-face interaction (non-IT tools) or groupware/netmeeting (IT tools).

The KM sub-processes used for SeLEKT are as follows:

- *Locating and accessing knowledge*: this sub-process involves the search and retrieval activities as well as guides to the location of knowledge. This includes locating the knowledge source, e.g. personal knowledge (in people's heads) or shared knowledge (in knowledge bases, in software systems, on the web).

- *Capturing knowledge*: this sub-process facilitates the codification of tacit knowledge in order to be more explicit and is one of the hardest sub-processes of KM because it is very difficult to capture people's knowledge and describe it without losing its context. This activity aims at capturing knowledge into systems, e.g. document, knowledge base, software or even videotape. It overlaps, in many cases, with the sub-process of representing knowledge.

- *Representing knowledge*: captured knowledge can be represented in different ways. The main aim of knowledge representation is to facilitate its access and transfer. Representation is facilitated by several means such as web publishing, video clips, documents, drawings and spreadsheets.

- *Sharing knowledge*: sharing knowledge focuses on the process of making knowledge sources available to users. The sharing of knowledge can take several forms. Face-to-face interaction is an important method of sharing tacit knowledge. The sharing of tacit knowledge can also be facilitated by some IT tools such as groupware and multi-media tools. The sharing of explicit knowledge also takes several forms. An individual or a group can access a knowledge base and obtain the stored knowledge.

- *Creating new knowledge*: the creation of new knowledge is one of the main objectives of KM. KM facilitates creating new knowledge from existing knowledge in several ways. Knowledge maps help in identifying the way different knowledge sources in an organisation relate to one another. Identifying the relationships helps in improving them and hence creating new knowledge. Similarly, data mining tools help in identifying the hidden relationships between data and hence creating new knowledge.

Table 6.7 presents the KM sub-processes and their 'technology categories'. It also includes a column listing some of the leading software applications that support every technology category. The use of this column will be discussed in stage three.

6.4.3   *Identify commercial software applications for the technology categories (stage 3)*

After a technology category is selected, as in stage 2, it is necessary to select a suitable software application. The selection of such applications is dependent on several factors such as the functional capabilities of the individual applications, the existing applications within the organisation and the ability to link them to the selected applications, the cost of the selected application, etc. A detailed description of the applications and associated characteristics has been developed in the SeLEKT approach.

6.5   Conclusions

This chapter has discussed the role of IT and non-IT tools in implementing a KM strategy. Existing approaches that are currently used to select KM tools are outlined and their limitations have provided the context for the development of a new approach, SeLEKT. This alternative provides a more informed approach in the selection of appropriate KM tools by incorporating three dimensions reflecting the organisational and knowledge context to ensure that the business needs of an organisation are adequately addressed. The approach described requires that an organisation understands its current and required status with regards to the three KM dimensions.

Table 6.7 KM sub-processes, supporting technology categories and software applications

KM sub-process	Technology tools	Commercial software applications
Locating and accessing	Experts directory	AskMe, Sigma Connect, IntellectExchange, Expertise Infrastructure
	Data warehouses	Syncsort: http://www.syncsort.com
	Web crawler – meta search	MetaCrawler, SurfWax, Copernic Basic 2001, Livelink, Dogpile, Mamma, CNET Search
	Data and text mining	Data Mining: Knowledge SEEKER, RetrievalWare, XpertRule Miner, Clementine
		Text Mining: SemioMap, Intelligent Miner for Text, Megapture Intelligence
	Knowledge mapping – concept mapping	Knowledge Service, IHMC Concept Map
	Intranet/extranet	Livelink, Instant Intranet Builder, iLevel
	Search engines	Google, Yahoo, FAST, Excite, AltaVista, Infoseek
	Taxonomy/ontological tools	Autonomy, SemioMap, RetrievalWare Suite
	Web mapping tools	Web Squirrel, WINCITE
	Electronic document management systems	Documentum, BASIS, Dicom
	Electronic mail	Eudora, Microsoft Outlook
Capturing	Word processors	MS Word, Word Perfect
	Case-based reasoning – expert systems	CBR-Works, Kaidara
	Knowledge bases	Assistum, KnowledgeBase.net, XpertRule Knowledge Builder
	Knowledge mapping – concept mapping	Knowledge Service, IHMC Concept Map
Representing	Mind mapping applications – brainstorming	Mind Manager, The Brain
	Web publishing	KnowledgeBase.net
	Virtual reality tools	Maelstrom, 3ds max for Windows
	Word processors	MS Word, Word Perfect
	Computer-aided design	Autodesk products
	Spreadsheets	MS Excel, StarOffice/OpenOffice Calc, Lotus 1-2-3
	Knowledge mapping – concept mapping	Knowledge Service, IHMC Concept Map
Sharing	Web publishing	KnowledgeBase.net
	Communities of practice	AskMe
	Intranet/extranet	Livelink, Instant Intranet Builder, iLevel
	Web-based file sharing tools	KnowledgeDisk, Briefcase
	Instant messaging	NetLert 3 Messenger, Trusted Messenger, ICQ, AOL Instant Messenger, Yahoo Messenger, MSN Messenger
	Integrated groupware solutions	A group of Lotus products (Notes, Domino, Sametime, QuickPlace), GroupWise, BrightSuite Enterprise, MyLivelink, Plumtree Collaboration Server, iTeam, iCohere
	Multi-media tools – video conferencing software	MS NetMeeting, AbsoluteBUSY, eRoom, WebEx Training Center, WebEx Meeting Center, WebDemo
	Electronic mail	Eudora, MS Outlook

Table 6.7   (*continued*)

KM sub-process	Technology tools	Commercial software applications
Creating	Data and text mining	Data Mining: Knowledge SEEKER, RetrievalWare, XpertRule Miner, Clementine Text Mining: SemioMap, Intelligent Miner for Text, Megapture Intelligence
	Mind mapping applications/ brainstorming	Mind Manager, The Brain
	Knowledge mapping – concept mapping	Knowledge Service, IHMC Concept Map
	Data warehouses	Syncsort

References

Al-Ghassani, A.M., Robinson, H.S., Carrillo, P.M. and Anumba, C.J. (2002) A framework for selecting knowledge management tools. *Proceedings of the 3rd European Conference on Knowledge Management (ECKM 2002)*, Trinity College, Dublin, Ireland, 24–25 Sept, pp. 37–48.

Anumba, C.J., Bloomfield, D., Faraj, I. and Jarvis, P. (2000) *Managing and Exploiting Your Knowledge Assets: Knowledge Based Decision Support Techniques for the Construction Industry, BR382*. Construction Research Communications Ltd, London.

Bair, J.H. and O'Connor, E. (1998) The state of the product in knowledge management. *Journal of Knowledge Management*, **2**(2), 20–27.

Barth, S. (2000) The power of one. *Knowledge Management Magazine*, CurtCo Freedom Group, **3**(12), 30–36.

Carrillo, P.M., Robinson, H.S., Al-Ghassani, A.M. and Anumba, C.J. (2004) Knowledge management in construction: drivers, resources and barriers. *Project Management Journal*, **35**(1), 46–56.

Davenport, T.H. and Prusak, L. (1998) *Working Knowledge: How Organisations Manage What They Know*. Harvard Business School Press, Boston, Massachusetts.

Egbu, C.O. (2000) The role of IT in strategic knowledge management and its potential in the construction industry. *Proceedings of the UK National Conference on Objects and Integration for Architecture, Engineering, and Construction*, 13–14 March, BRE, Watford.

Gallupe, R.B. (2001) Knowledge management systems: surveying the landscape. *International Journal of Management Reviews*, **3**(1), 61–7.

Haag, S. and Keen, P. (1996) *Information Technology, Towards Advantage Today*. McGraw-Hill, New York.

Jackson, C. (1998) Process to product: creating tools for knowledge management. *Proceedings of the 2nd International Conference on Technology Policy and Innovation, Assessment, Commercialisation and Application of Science and Technology and the Management of Knowledge*, 3–5 Aug, Lisbon, Portugal.

Lang, J.C. (2001) Managerial concerns in knowledge management. *Journal of Knowledge Management*, **1**(1), 43–57.

Laudon, K.C. and Laudon, P.L. (2000) *Management Information Systems*, 6th edition. Prentice-Hall, New Jersey, p. 436.

Lucca, J., Sharda, R. and Weiser, M. (2000) Coordinating technologies for knowledge management in virtual organisations. *Proceedings of the Academia/ Industry Working Conference on Research Challenges CAIWORC'00*, 27–29 April, Buffalo, New York.

Nonaka, I. (1995) *The Knowledge-Creating Company: How Japanese Companies Create the Dynamics of Innovation*. Oxford University Press, New York.

Ruggles, R. (1997) *Knowledge tools: using technology to manage knowledge better*. Working paper for Ernst and Young, <http://www.businessinnovation. ey.com/mko/html/toolsrr.html> (26 August 2000).

Storey, J. and Barnet, E. (2000) Knowledge management initiatives: learning from failure. *Journal of Knowledge Management*, **4**(2), 145–56.

Tiwana, A. (2000) *The Knowledge Management Toolkit*. Prentice-Hall, New Jersey.

Tsui, E. (2002a) Technologies for personal and peer to peer (P2P) knowledge management. *CSC Leading Edge Forum Technology Grant Report*, <http:// www2.csc.com/lef/programs/grants/finalpapers/tsui_final_P2PKM.pdf> (06 June 2002).

Tsui, E. (2002b) Tracking the role and evolution of commercial knowledge management software. In: *Handbook on Knowledge Management* (C.W. Holsapple, ed.). Springer, Berlin/Heidelberg.

Wensley, A.K.P. and Verwijk-O'Sullivan, A. (2000) *Tools for knowledge management*. Internal report, Joseph L. Rotman School of Management, University of Toronto, Ontario, <http://www.icasit.org/km/toolsforkm.htm> (25 June 2002).

7 Cross-Project Knowledge Management

John M. Kamara, Chimay J. Anumba
and Patricia M. Carrillo

7.1 Introduction

A project can be defined as a 'unique process, consisting of a set of co-ordinated and controlled activities with start and finish dates, undertaken to achieve an objective conforming to specific requirements including constraints of time, cost and resources' (ISO8402 in Lockyer and Gordon, 1996, p. 1). Another definition of a project is that of 'a one-time, multitask job with a definite starting point, definite ending point, a clearly defined scope of work, a budget and usually a temporary team' (Lewis, 2001, p. 5). These definitions suggest a number of characteristics of projects which include the following:

- uniqueness ('one-time' activity) and determinate life
- the temporary nature of project organisations, usually involving a multi-disciplinary (and possibly multi-organisation) team
- the goal of a project is to achieve 'something'
- set of (complex) interrelated activities to achieve the overall goal.

The uniqueness and goal-oriented nature of projects imply that they are insular self-contained activities; but it can also mean that within a project, there can be several smaller 'projects'. For example, a work package or specific phase in a large project can itself constitute a 'project', especially where a multi-organisation (or multi-department) team is delivering the project. The determinate nature of projects helps to focus resources in accomplishing a specific goal (or set of objectives), but it should be understood that projects do not exist in isolation; they are usually set within the context of one or more 'permanent' (or established and 'ongoing') organisations.

Although projects comprise a set of interrelated activities, they usually have definable phases, for example conception, development, realisation and termination (Lockyer and Gordon, 1996). At the conception stage the need for the project is established and broad goals (e.g. budget, time frame, etc.) are set. This phase is about the conceptualisation of ideas and clarification of requirements for the project. The development stage involves detailed project design, specification and planning. The realisation phase involves actual implementation of the project according to the plans and specifications developed in the previous phase. The termination phase deals with reviewing the project and collating lessons learnt.

An important point to note about project phases is that of the changing nature of the project organisation. The earlier phases are more dynamic as they involve the generation/conceptualisation of ideas (an 'organic' type of organisation). The later stages (especially the realisation phase) are much more structured as they involve the implementation of plans (a more 'mechanistic' organisation). Thus, it can be said that as a project develops, its activities and tasks become more specific, and the project organisation becomes more structured.

This chapter discusses the issues involved in the management of construction project knowledge. Following the Introduction, the next section (7.2) outlines the nature of projects, highlighting the relationship between projects and parent organisations, and the implications for knowledge management (KM). Section 7.3 focuses on specific issues relating to the characteristics of construction projects, the management of construction project knowledge in terms of content (knowledge to be managed) and context (organisation in which it is managed), and the knowledge requirements at different stages of a construction project. Section 7.4 examines the issue of knowledge transfer across different projects, its dimensions in terms of role of individuals, project reviews, contractual and organisational arrangements and the goal of knowledge transfer. Section 7.5 discusses various approaches for live capture and reuse of project knowledge, distinguishing between 'soft' concepts and 'hard' technologies. A conceptual framework for the live capture and reuse of project knowledge is described as a way forward for successful cross-project KM in the construction industry. The chapter concludes (Section 7.6) that live capture reflects a step change in the way construction project knowledge is managed, but further research is needed in this area to improve the efficiency of project delivery.

7.2   The nature of projects

The nature and characteristics of projects impinge on how knowledge is managed within and between projects. The changing nature of the project organisation, for example, has implications on KM across the lifecycle of a project. According to Ståhle (1999), the organic part of an organisation is more amenable to people-centred KM strategies; the mechanistic part, which deals more with explicit knowledge, can rely on structured procedures and IT tools. The effective management of knowledge within a project therefore needs to address these changing organisational and KM requirements.

The 'context' within which a project is implemented also provides the basis for the transfer of knowledge, either between the project and the parent organisation, or between multiple projects, or between multiple projects and the parent organisation. These interactions are illustrated in Figure 7.1. Figure 7.1a shows a situation where a 'parent organisation' has a one-off project (e.g. the installation of a new software system). The existing knowledge base within the organisation is used to set up the project; after implementation, the learning from the project is fed back into the corporate knowledge base (or is lost). Figure 7.1b shows a parent organisation with several projects (running at the same time or sequentially). In this situation there is, supposedly, a cyclical transfer and reuse of knowledge between the organisational knowledge base (OKB) and projects. Where projects are running in parallel, there can be transfer between projects; if sequential, then transfer between projects is via the OKB.

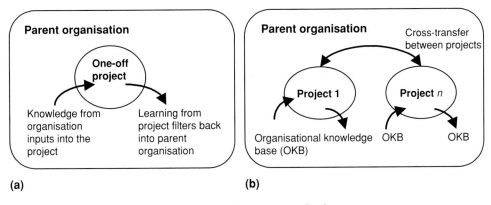

(a)   (b)

Figure 7.1   Relationships between projects and parent organisations

7.3   Construction projects

The construction industry (otherwise referred to as the architecture/ engineering/construction (AEC) industry) is a project-based industry. Its business is organised around projects, which are delivered by multidisciplinary organisations working together in temporary 'virtual' organisations. The goal of these projects is the creation, refurbishment and/or demolition of a built asset (e.g. building, road, bridge).

There are different 'levels' of projects within the construction industry. There is the 'main project' (hereinafter called the 'construction project'), which is commissioned by a client and implemented by multi-disciplinary and multi-organisation teams. The portion (or work package) of the construction project (e.g. design of facility) implemented by participating firms is also a (mini) project from the perspective of the contributing organisation. Thus an organisation (e.g. architectural firm) can have many 'projects', which are themselves parts of different construction projects. From the client's perspective, the construction project is part of their business operations (or part of a wider project such as the restructuring of a particular business unit). Depending on the nature of these business operations, there can be several such 'construction projects'. There are therefore different linkages and perspectives from which to view construction projects. These can be summarised in the following statements:

(1) A {Construction Project (CP)} is a subset of {Client Business Operations/Project}
(2) A {Construction Organisation Project (COP)} is a subset of {CP}
(3) A {COP} is a subset of {Construction Organisation Project Portfolio (COPP)}

These linkages are illustrated in Figure 7.2. The focus in this chapter is on the construction project (CP), particularly from the perspective of the firms (consultants, contractors, etc.) that make up the construction industry (i.e. the supply side that provide facilities for clients).

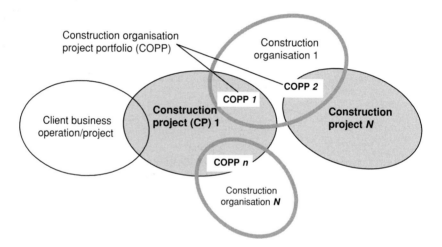

Figure 7.2   Linkages between projects in the construction industry

Like other projects, construction projects have definable phases for which there are many configurations. For simplicity, these phases are: conception, design and specification, construction, commissioning and handover. The conception stage is mostly within the client organisation although it is increasingly being recommended that construction professionals are involved during this phase. It deals with all the issues/decisions about the need for the project, how it relates to the client's business need, and a specification of broad project goals. Key members (e.g. project manager, lead designer, etc.) of the project team are also appointed at this stage. An outcome of this phase is a set of client requirements (or a project brief) which, depending on the project, can have varying levels of detail about the wishes of the client, site and other factors likely to affect the project.

During the design and specification phase, the brief is 'translated' into a design of the facility with detailed specifications on materials, cost and, in some cases, work schedules. The construction phase deals with actual 'building' on site, but can also be preceded by a detailed plan of how this process will be carried out. Commissioning and handover deals with the transition from suppliers (i.e. construction firms delivering the project) back to the client who is then expected to use, operate and maintain the facility. However, in some cases, e.g. in private finance initiatives (PFIs), the remit of construction firms includes the operation of the facility over a specified number of years; there is therefore no immediate 'handover' to the client (or owner).

It should be pointed out that project phases (especially conception and design/specification) are not entirely linear as there is a lot of iteration in the early stages. This also suggests a change in the nature of the project organisation – from 'organic' to 'mechanistic' – with implications for managing project knowledge.

7.3.1 Management of construction project knowledge

The successful management of construction project knowledge (like KM in general) depends on the knowledge to be managed (*content*) and the *context* (organisation) in which it is managed (Kamara *et al.*, 2002). Issues relating to project organisation have been discussed above, but these will be revisited later.

Project knowledge is defined here as: knowledge (including data and information) required to conceive, develop, realise and terminate a project. This definition reflects the view by Blumentritt and Johnston (1999) that knowledge is a component of a task-performing system; that is, a state of the system that warrants task completion and the future repetition of this task. Within the construction domain, project knowledge is interconnected and includes knowledge about the end product (i.e. the facility being constructed), the processes involved in its creation and the resources needed. Knowledge requirements therefore include issues such as the nature and wishes of the client and end-users of the facility (client/use requirements), the nature of the site on which the facility is to be built and the immediate environment (site information), and information on the regulations that apply to that facility (e.g. building, planning, and health and safety regulations). Knowledge about the end-product that will satisfy the competing requirements and constraints is also necessary. The whole process involves the collection, processing (or conversion) and representation of knowledge (e.g. converting the wishes of the client into a design) to enable the erection of a facility that would satisfy the client. The knowledge requirements with respect to the different stages in a construction project and the professionals involved are summarised in Table 7.1.

Table 7.1   Knowledge requirements and professionals involved in a construction project

Project stage	Knowledge requirements	Professionals involved (usually from different firms)
Conception	Do we need a project? What purpose will the facility serve? How much funds can we commit to the project? Where is it going to be built? How will the facility be procured? Who are the best firms to do the job? Who will look after our interests?	Development managers; property consultants; project managers; financial consultants; facility managers; business managers; planning authorities
Design and specification	What does the client really want? What are the characteristics of the site for the facility? How do the relevant regulations apply to this facility? What will the facility look like? What kind of materials do we need, and in what quantity? How much is it going to cost? How will the facility be constructed?	Architects; engineers (building services, civil, structural); planning supervisors; facility managers; quantity surveyors; project managers; contractors
Construction of facility	Have all materials been specified? Where can we obtain the materials? How and when do we need them on site? How can we organise ourselves better to do the job efficiently? How can different components be assembled efficiently? How can we ensure the quality of workmanship?	Contractors; project managers; specialist contractors; materials and equipment suppliers; architects; engineers
Commissioning and handover	Is the facility performing as expected? Is it serving the purpose for which it was created? Are all components and systems working effectively? Are all interest groups satisfied?	Facility managers (operators); architects; contractors; client/end users

The knowledge required to deliver construction projects is fragmented; it is held by different professionals who are based in separate firms. This reflects the nature of the project organisation, which is essentially a temporary multi-disciplinary organisation. Because of these disparate repositories of knowledge, a key aspect of project KM in construction is therefore the transfer or mutual sharing of knowledge for the 'common good' of the project at different levels:

(1) The transfer/sharing of knowledge between different professionals involved in each phase of the project (e.g. between architects and engineers involved in the design of a building).

(2) The transfer/sharing of knowledge between different stages of the project (e.g. from the client's business need at conception to design/specification of the facility that satisfies that need).

(3)   The mutual transfer of knowledge from a project to the organisational knowledge base (OKB) of each firm involved in a construction project (see Figures 7.1 and 7.2) (i.e. the transfer of knowledge between projects, via a particular construction firm).

The first two levels of transfer depend on the kind of organisation (i.e. procurement, contractual and communications) set up to manage the project, but also on the third level of transfer. It is at this level (between project and participating firm) that learning from a project is stored and reused on subsequent projects – the knowledge transfer at level 3 is used to support transfers at levels 1 and 2. Individual construction firms are therefore key to the effective transfer (and hence, management) of project knowledge.

7.4   Cross-project knowledge transfer

Brookes and Leseure (2000) define cross-project KM as 'a subset of knowledge management that focuses on the transfer of knowledge across different projects' using a variety of strategies depending on the organisational context (or sector) being considered. In some cases, the 'strategies' used occur by default. For example, in the manufacturing industry, studies on KM reported in Brookes and Leseure (2000) suggest that cross-project KM is not explicitly undertaken, although companies can identify business benefits that would arise from better cross-project KM. However, despite the dearth of formal and explicit processes, significant amounts of knowledge are transferred between projects both in tacit and explicit forms. Knowledge was transferred through recruitment, mentoring, informal organisational networks, databases, intranets and expert systems. It was further established that, to be successful at cross-project KM, companies needed to identify the high-grade knowledge and maintain this very carefully. High-grade knowledge should be made as explicit as possible. It should be well maintained and changed on the basis of experience. At the same time, companies must not neglect background knowledge, and they need to develop a self-sustaining infoculture to maintain this.

Successful transfer of knowledge between different projects is influenced by the way knowledge is captured (i.e. when and how) and repackaged (or codified) for reuse. Whatever process is set in place to achieve this should seek to do the following:

- Facilitate the reuse of the collective learning on a project by individual firms and teams involved in its delivery.
- Provide knowledge that can be utilised at the operational and maintenance stages of the asset's lifecycle.

- Involve members of the supply chain in a collaborative effort to capture learning in tandem with project implementation, irrespective of the contract type used to procure the project from the basis of both ongoing and post-project evaluation.

The issues of 'collective learning' and 'supply chain involvement' require concerted efforts to integrate the disparate stores of project knowledge for the mutual benefit of individual members (firms) in the project team. It poses a number of challenges, among which are the following:

- What knowledge from a project is reusable in other projects?
- How can this knowledge be captured (during and after project implementation) in a cost-effective way, given the temporary nature of construction projects, and given the various facets (e.g. organisational, human and technology issues) that need to be considered?
- How can project knowledge be captured without causing unnecessary knowledge overload for project participants who already have to cope with huge amounts of project information?
- In what ways can captured knowledge be made available for reuse during (and after) project execution?

The questions and issues raised above provide the background against which current practice for knowledge transfer on construction projects is discussed.

7.4.1 Knowledge transfer on construction projects

The transfer of knowledge across different construction projects is situated within a firm's overall strategy (or *de facto* practice) for KM. Studies on KM practice, reported in Kamara *et al.* (2002), suggest that the management of knowledge in construction firms include the following elements:

- 'A strong reliance on the knowledge accumulated by individuals, but there is no formal way of capturing and reusing much of this knowledge.'
- 'The use of long-standing (framework) agreements with suppliers to maintain continuity (and the reuse and transfer of knowledge) in the delivery of projects for a specific client.'
- 'The capture of lessons learnt and best practice in operational procedures, design guidelines, etc., which serve as a repository of process and technical knowledge. Post-Project Reviews (PPR) are usually the means for capturing lessons learned from projects.'
- 'The involvement (transfer) of people in different activities as the primary means by which knowledge is transferred and/or acquired.'

- 'The use of formal and informal feedback between providers and users of knowledge as a means to transfer learning/best practice, as well as to validate knowledge (for example, site visits by office-based staff to obtain feedback on work progress).'
- 'A strong reliance on informal networks and collaboration, and "know-who" to locate the repository of knowledge.'
- 'Within firms with hierarchical organisational structures, there was a reliance on departmental/divisional heads to disseminate knowledge shared at their level, to people within their sections.'
- 'The use of appropriate IT tools (such as GroupWare, Intranets) to support information sharing and communication.'

All the strategies listed above are directly and indirectly associated with the management of project knowledge, since the core business of construction firms is essentially projects-based. However, the heavy reliance on knowledge accumulated by individuals, project reviews (including post-project evaluation) and specific contractual/organisation arrangements (e.g. framework agreements) are considered to be the key approaches for direct transfer of project knowledge (Orange *et al.*, 1999).

Cross-project knowledge transfer by individuals

People are still considered to be the key resource of any organisation, and they play a vital role in cross-project knowledge transfer. The reliance on people is based on the assumption that the knowledge acquired from one project can be transferred by that individual when he/she is reassigned to another project. This strategy is usually reflected in job rotation and mentoring for junior staff to help them 'learn on the job'. The scenario is similar to that in Figure 7.1b (modified in Figure 7.3) except that, in this case, the transfer from each project is to individuals rather than to an organisational knowledge base (in the sense of it being corporately held in some 'central' repository).

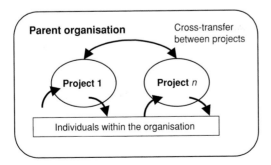

Figure 7.3   The role of individuals in the transfer of knowledge across projects

The effectiveness of knowledge capture and reuse by individuals within an organisation depends on the following:

- The level of involvement in the main project. If involvement is in some specific aspect/stage of the project, then learning will be confined to a limited perspective of the project. If, on the other hand, involvement is over the entire project, then there is potential for knowledge capture and reuse.
- The ability to consolidate the learning from a previous project. If there is inadequate time between projects, there will be little scope for effective knowledge transfer on subsequent projects.
- The length of time they stay with the organisation. If an individual leaves a firm, there is always a potential loss of knowledge, even if efforts have been made to mine that knowledge beforehand.

A key issue in a people-based knowledge transfer strategy is therefore the relationship between 'individual knowledge' and 'shared organisational knowledge'; that is, how much knowledge is retained individually, and how much is held corporately.

Project reviews

Project reviews facilitate the consolidation of learning and, to some extent, create shared understanding on a project. They include ongoing (phase) reviews – both formal and informal (e.g. design reviews) and post-project evaluation. One of the primary reasons for post-project evaluation cited by the Office of Government Commerce (OGC) is 'to transfer the knowledge and any lessons from one project to other projects' (OGC, 2002). There are various guides on how to carry out project reviews (e.g. OGC, 2002, for public sector clients in the UK), and some organisations have well-established procedures. The effectiveness of such reviews undoubtedly depends on the way they are conducted, but also on the time allocated for reviews and the availability of the relevant staff.

Contractual and organisational arrangements

The use of contractual and organisational arrangements for cross-project knowledge transfer is designed to influence the level of involvement of individuals (or firms) in the main project. For example, design-and-build contracts allow a contractor to be involved much earlier in the project and will have the continuity of knowledge transfer across different stages of a project; the learning on the project is also likely to be from a much wider perspective because of the extensive involvement. In private finance initiative (PFI) contracts where contractors are involved in financing, designing, constructing and maintaining a facility over an agreed period of time, there is even more scope for effective knowledge transfer across project stages and also over the lifecycle of a facility.

Another organisational arrangement that facilitates cross-project knowledge transfer is the framework agreement. This involves the creation of long-term (e.g. five-year) relationships with a group of suppliers (i.e. construction firms) to maintain continuity in the delivery of projects for a specific client with a continuous construction programme. Although there is usually no guaranteed work for each supplier in the framework agreement, the intention is that the learning by individuals and firms is reused on future projects.

It is important to note that the effectiveness of various contractual and/or organisational arrangements in facilitating cross-project knowledge transfer is dependent on whether the same people are used, or whether there is a strategy for sharing individual knowledge across the organisation.

7.4.2   Current practice v. goals for knowledge transfer

Earlier in this section, it was established that effective cross-project knowledge transfer is enhanced by 'collective learning' on a project and 'supply chain involvement'. Collective learning in this context implies the involvement of the supply chain (i.e. all the firms involved in delivering a project) in capturing the learning across the project. This means that firms not only have access to their small chunk of knowledge arising from their particular part of a project, but also can benefit from the wider learning gained on the whole project. However, current practice for cross-project knowledge transfer does restrict a participating organisation's learning to its particular perspective of the project. The organisation also has limitations which affect knowledge transfer not only on projects but also on the wider remit of KM.

For example, the reliance on people, even within a framework agreement, makes organisations vulnerable when there is a high staff turnover. The use of framework agreements also cannot guarantee that the learning of individual firms participating in the agreement is shared with other participants for the benefit of the project, since these firms can be in competition elsewhere (e.g. on other projects) and may not want to divulge 'secrets' that might weaken their competitive advantage. Project reviews (especially post-project reviews) are useful in consolidating the learning of the people involved in a project, but they are not very effective in transferring knowledge to non-project participants. It is arguable whether reviews that are conducted during projects sufficiently focus on the capturing and reuse of learning, or whether the focus is on achieving project targets.

Improving cross-project knowledge transfer can be done by implementing the strategies discussed in Chapter 10 or by devising ways of capturing reusable knowledge *during* project implementation and presenting it in a format that can facilitate its reuse during and after the project. A conceptual framework for achieving this is described next.

7.5   Live capture and reuse of project knowledge

The development of an appropriate methodology for the live capture of construction project knowledge involves the use of both 'soft' (i.e. organisational, cultural and people issues) and 'hard' (information and communication technologies, ICTs) concepts and tools. Lessons from past research initiatives suggest that the combined approach of 'soft' and 'hard' is the most sensible one to adopt. Previous work on KM focused on the delivery of technological solutions (Carrillo *et al.*, 2000), probably because of the growth in knowledge-based expert systems in the 1980s and early 1990s. However, it is now recognised that good KM does not result from the implementation of information systems alone (Davenport, 1997; Stewart, 1997). Therefore organisational and people issues, which are not readily solved by IT systems, would also need to be resolved (Tiwana, 2000). On the other hand, approaches that exclusively focus on organisational and cultural issues would not reap the benefits derived from the use of IT, especially in the context of distributed teams that are the norm in construction (Anumba *et al.*, 2000). For example, in the B-HIVE (Building a High Value Construction Environment) research project, which developed a cross-organisational learning approach (COLA) for learning and knowledge generation through reflection and discussion within a partnering context, it was concluded that the use of IT (e.g. a project extranet) would enhance the usefulness of the COLA system (Orange *et al.*, 1999). Thus a combined approach will deliver a more complete solution that incorporates 'soft' issues with 'hard' technological issues.

7.5.1   'Soft' concepts for live knowledge capture

These include the existing concepts of 'collaborative learning (CL)' and 'learning histories (LH)', which will be adopted for use within a construction project context. These concepts will play a key part in the development of the 'methodology' for live knowledge capture.

Collaborative learning is a business practice that is aimed at discovering explicit and tacit collaboration tools, processes and knowledge, experimenting with them and creating new knowledge from them (Digenti, 1999). It employs methods and approaches that emerge from the present situation and allow organisations to move across boundaries fluidly and to ensure that the learning that takes place in one group (e.g. a project) is transferred back to the organisation. This is quite appropriate to the context of construction projects, where a network of organisations is involved in delivering projects. This concept can therefore be adapted to facilitate the transfer and reuse of collective learning to the individual firms involved in implementing it.

A learning history is a process, originally developed at the Massachusetts Institute of Technology (MIT), for capturing usable knowledge from an extended experience of a team and transferring that knowledge to another team that may operate in a different context (Kleiner and Roth, 1997; Dixon, 2000). Construction projects and the teams that implement them are unique, but the structures of teams, processes, tools, skills, etc. used in these projects are similar, and provide the opportunity for the reuse of knowledge. Therefore, using the concept of a 'learning history', the learning of one team (from critical events on a project) can serve as a catalyst to a similar team in order to deal with issues in a different context.

7.5.2   'Hard' technologies for live knowledge capture

The 'hard' tools for knowledge capture include available ICT applications that are currently being used in the construction industry, particularly project extranets, workflow management tools and other groupware applications for collaborative working.

Project extranets (or project websites) are 'dedicated web hosted "collaboration and information spaces" for the AEC [architecture, engineering and construction] industry that support design and construction teams' (Augenbroe *et al.*, 2001). Extranets are hosted by a growing number of third-party organisations (application service providers, ASPs) which include Architec Ltd, 4Projects, Bidcom, BIW Technologies, Buzzsaw and BuildOnline (Kamara and Anumba, 2002). Project extranets utilise client-server technology and a web browser is usually all that is required to allow distributed project team members to share, view and comment on project-relevant information. The growing use of extranets in the delivery of construction projects, and the collaborative facilities they provide (CPN, 2002), make them suitable as a platform on which a methodology for live knowledge capture can be mounted. The proposed use of web-based technology will address the constraints of distributed teams, and would facilitate contributions from various members of a project without the need to meet in one location. Furthermore, the widespread availability and use of the internet would make it possible for most firms in the supply chain to be able to use the tool. However, because of the current limitations of extranets (e.g. being purely document-centric with limited facilities for workflow), other tools and technologies (e.g. workflow modelling and automation tools, web-server push technologies, etc.) will be utilised.

7.5.3   Conceptual framework for live knowledge capture

The real-time capture of project knowledge can be effected through the following: a project knowledge file, an integrated workflow system and a project knowledge manager. However, before these components are described, an overview of procedure for knowledge capture will be presented.

Overview of the knowledge capture procedure

Figure 7.4 shows an overview of the knowledge capture procedure. During the course of a construction project, learning occurs not only from many critical events but also from the normal day-to-day operations. This learning can be about the facility being constructed, the project process or the participants involved in its execution. Within the knowledge capture procedure, the structure of how learning is captured is determined beforehand in the project knowledge file. When a learning event occurs, the integrated workflow system is triggered and this sets in motion a flow of actions to capture the learning at a particular point in time. The learning is compiled and edited for reuse within the current project, or in subsequent projects.

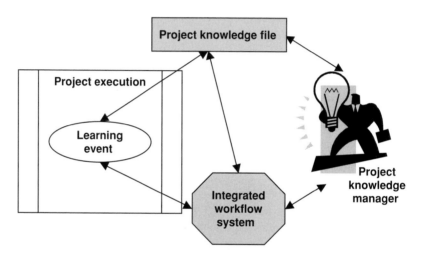

Figure 7.4   Overview of the knowledge capture system

Project knowledge file

The 'project knowledge file' (PKF) is similar to the Health and Safety File (HSF) under the Construction (Design and Management) (CDM) Regulations in the UK. 'The health and safety file is a record of information for the client which focuses on health and safety' (HSE, 1997). The detailed information collected depends on the nature and size of the project, but it is used to 'alert those who are responsible for the structure of the key health and safety risks that will need to be dealt with during subsequent maintenance, repair and construction work' (HSE, 1997).

Similar to the HSF, the PKF will contain information relating to the 'project knowledge', but will focus on knowledge that can be reused both during the execution (e.g. in subsequent phases) and after the completion of the project. The kind of knowledge to be captured and the format and contents of the PKF will be determined through detailed research into reusable project knowledge, but the goal will be to develop an ongoing 'learning history' for the project within a collaborative environment. The PKF is agreed on at the onset of a project and all parties are required to contribute to its compilation. The PKF should be managed by a designated project knowledge manager (similar to the role of a planning supervisor under the CDM Regulations), but depending on the size and nature of the project, this role can be combined with that of the project manager.

Integrated workflow system

The role of the integrated workflow system (IWS) is to implement the PKF in real-time. That is, to facilitate the compilation of the learning history for the project during its execution, in accordance with the parameters set out in the PKF. A generic model for the workflow will be developed following research into the format and contents of the PKF, but it should be customisable to take into account variations in the PKF. The IWS is triggered when a 'learning event' (i.e. an activity or event from which reusable knowledge can be captured) takes place. This can be, for example, problems and how they were solved, innovations, breakthroughs or the normal day-to-day operations of a project. When such events occur, the IWS will request the relevant participants (in accordance with the agreement set out in the PKF) to contribute their views on how various issues were dealt with. A compilation of these different perspectives will form part of the 'learning history' at particular stages of a project, which can be reused either at subsequent phases or at the end of a project. The trigger for the IWS can be either automated, done manually by a project knowledge manager, or a combination of both manual and automated systems. An automated trigger requires data and text-mining capabilities or other means of detecting, say within a project extranet, when certain events occur. In both automated and manual triggers, server 'push' technologies will be utilised to ensure that the required prompt is pushed to relevant participants. The IWS should also have filtering capabilities to ensure that only the relevant learning is captured to prevent knowledge/information overload. The IWS can be integrated with existing project extranets or can be developed as a separate application that is compatible with extranets.

Project knowledge manager

This is a role (as opposed to a particular individual) that will be charged with developing and managing the PKF and the IWS. A person or persons occupying this role will need to be familiar with the principle of learning histories and how they are developed.

Potential benefits

The live knowledge capture system, through ensuring the *currency* and *relevance* of the knowledge captured, will have a significant impact on the overall construction process:

(1) Construction supply chains will benefit through the shared experiences that are captured as part of the learning on key events (e.g. problems, breakthroughs, change orders, etc.). The benefits to this group are both short and long term: short term in the sense that project teams would be enabled to manage better the subsequent phases of a project (through the capture and transfer of learning from a previous phase); long term because it will increase their capacity to better plan future projects and their ability to collaborate better with other organisations. Furthermore, learning from past projects can be used to train new employees and project managers.

(2) Other project teams can use the learning captured from previous/similar projects to deal with problems; reflection on previous learning can also trigger innovative thinking (to think about issues that might be relevant to their project).

(3) Client organisations will benefit from enriched knowledge about the development and construction of their assets. This will contribute to the effective management of facilities and the commissioning of other projects. In the longer term, clients will benefit from the increased certainty with which construction firms can predict project outcomes.

(4) Project staff and students of project/construction management and the institutions providing such courses/training will also benefit through the use of captured project knowledge as case study material.

(5) It will lead to improved supply chain management, as team members would work more collaboratively and share the lessons learnt on construction projects.

The construction industry will benefit from an enhanced knowledge base as much learning that is presently not documented can be captured and reused.

7.6   Conclusions

This chapter has discussed various issues for cross-project KM, with particular focus on the organisational implications for achieving this. Current practices for the transfer of project knowledge have also been considered and their limitations highlighted. The proposed framework for live capture and reuse of project knowledge constitutes a step change in KM in construction and its implementation, which requires further research and development. This will ensure the currency and relevance of project knowledge and will have significant impact on the efficiency of project delivery in the construction industry.

References

Anumba, C.J., Bloomfield, D., Faraj, I. and Jarvis, P. (2000) *Managing and Exploiting Your Knowledge Assets: Knowledge Based Decision Support Techniques for the Construction Industry.* Construction Research Communications Ltd, London.

Augenbroe, G., Schwarzmueller, G. and Verheij, H. (2001) Project web sites with project management extensions: a prototype. In: *Design Management in the Architectural and Engineering Office* (G. Augenbroe and M. Prins, eds), pp. 395–404. CIB, Rotterdam.

Blumentritt, R. and Johnston, R. (1999) Towards a strategy for knowledge management. *Technology Analysis and Strategic Management,* **11**(3), 287–300.

Brookes, N.J. and Leseure, M. (2000) *Cross-Project Knowledge Management: The State of the Art.* Internal Report, CLEVER Project, Loughborough University.

Carrillo, P.M., Anumba, C.J. and Kamara, J.M. (2000) Knowledge management for construction: key IT and contextual issues. In *Proceedings of the International Conference on Construction IT* (G. Gudnason, ed.), 28–30 June, Reykjavik, Iceland, Icelandic Building Research Institute, pp. 155–65.

CPN (2002) *Project Extranets – The Real Benefits.* Construction Productivity Network (CPN), London (available at http://www.ciria.org.uk/cpnww/0001.htm) (accessed 17 Feb 2004).

Davenport, T. (1997) Secrets of successful knowledge management. *Knowledge Inc,* **2**, February.

Digenti, D. (1999) *The Collaborative Learning Guidebook.* Learning Mastery, Somerville, Massachusetts.

Dixon, N.M. (2000) *Common Knowledge.* Harvard Business School Press, Boston, Massachusetts.

Health and Safety Executive (HSE) (1997) *Construction (Design and Management) Regulations 1994: The Health and Safety File.* HSE Information Sheet, Construction Sheet No. 44 (available at http://www.hse.gov.uk/pubns/con44.htm) (accessed 6 Jan 2003).

Kamara, J.M. and Anumba, C.J. (2002) Collaborative systems and CE implementation in construction. In: *Proceedings of the 3rd International Conference on Concurrent Engineering in Construction* (I.D. Tommelein, ed.), Berkeley, California, 1–2 July, pp. 87–98.

Kamara, J.M., Augenbroe, G., Anumba, C.J. and Carrillo, P.M. (2002) Knowledge management in the architecture, engineering and construction industry. *Construction Innovation,* **2**, 53–67.

Kleiner, A. and Roth, G. (1997) How to make experience your company's best teacher. *Harvard Business Review,* **75**(5), 172–7.

Lewis, J.P. (2001) *Project Planning, Scheduling, and Control,* 3rd edition. McGraw-Hill, New York.

Lockyer, K. and Gordon, J. (1996) *Project Management and Project Network Techniques,* 6th edition. Pitman Publishing, London.

Office of Government Commerce (OGC) (2002) *Project Evaluation and Feedback* (available from the OGC website: http://www.ogc.gov.uk/sdtoolkit/ reference/achieving/achievin.html#lib9 (accessed June 2003).

Orange, G., Burke, A. and Cushman, M. (1999) An approach to support reflection and organisational learning within the UK construction industry. *Paper presented at BITWorld'99*, Cape Town, 30 June–2 July (http://is.lse.ac.uk/b-hive).

Ståhle, P. (1999) New challenges for knowledge management. In: *Liberating Knowledge* (J. Reeves, ed.), pp. 36–42. Caspian, London.

Stewart, T.A. (1997) *Intellectual Capital: The New Wealth of Organisations.* Doubleday, New York.

Tiwana, A. (2000) *The Knowledge Management Toolkit.* Prentice-Hall, New Jersey.

8 Knowledge Management as a Driver for Innovation

Charles O. Egbu

8.1 Introduction

There is general agreement among researchers, practitioners and governments that a relationship exists between an organisation's efficiency or profitability and its ability to innovate (Gann, 2000; Egbu, 1999a, 1999b, 2001a). Innovation is viewed as a major source of competitive advantage and is perceived to be a prerequisite for organisational success and survival (Egbu, 1999a; Gann, 2000; Egbu and Botterill, 2001).

Innovation is defined as the 'successful exploitation of an idea, where the idea is new to the unit of adoption' (Egbu, 2001a, p. 1). Creativity is only the generation of the idea itself. Innovations come from many different sources and exist in many different forms (Tatum, 1987; Slaughter, 1993; Wolfe, 1994). There is a dichotomy between radical and incremental innovation. Innovation can be radical, in response to crisis or pressure from the external environment, but it can also be incremental, where step-by-step changes are more common. Moreover, a common typology distinguishes product and process innovation. Product innovation describes where a new product is the outcome. It is seen to focus on cost reduction by obtaining a greater volume of output for a given input. Process innovation denotes innovation where the process by which a product is developed is exposed to new ideas and, therefore, leads to new, often more sophisticated methods of production. It describes new knowledge, which allows the production of quality superior output from a given resource. From the perspective of sources of organisational innovations, there are emergent (from within), adapted/adopted and imposed innovations.

Innovation is often not simple or linear, but is, rather, a complex iterative process having many feedback and feed-forward cycles. In essence, innovation can be viewed as a process of interlinking sequences from idea generation to idea exploitation that are not bound by definitional margins and are subject to change. Therefore, it is necessary to understand the complex mechanisms of this process and the context in which the innovation takes place (Wolfe, 1994).

There is still an ongoing debate on whether the construction industry and many of the construction organisations are innovative or not. There are those who suggest that the construction industry is less innovative than many other industries. The author posits the view that many of these debates are founded on weak premises. Understanding the innovation trajectory that an organisation embarks upon gives a better understanding as to whether the organisation has been successful at innovating or not. Industrial sectors and organisations are impacted upon by different constraints and they handle these differently. Organisations innovate for so many reasons. There are also different drivers that fuel innovation. Organisations might innovate to increase profit share, to enter a new market, to be a leader or first follower in the market, or for reasons of status. Organisational strategies for innovation differ from one organisation to another. Similarly, the approaches that organisations put forward for measuring their innovation success as well as the time frame for judging innovation success differ greatly. What is perceived to be a highly successful innovation for one organisation may not be seen to be so by another organisation. There are organisations that choose to exnovate (cease to continue with their innovation) say three or five years after the release of their innovative products or solutions. There are some that might measure the success of their innovation after ten to fifteen years. It is therefore important to understand the *modus operandi* of an organisation involved in innovation before the judgement is made whether the organisation is successful at innovation or not.

The innovation strategies of organisations are, however, strongly constrained by their current position and core competencies as well as the specific opportunities open to them in future. In other words, organisational strategies for innovation are 'path-dependent'.

The ability to innovate depends largely on the way in which an organisation uses and exploits the resources available to it. A vital organisational resource, at the heart of innovation, is knowledge. Knowledge is fast overtaking capital and labour as the key economic resource in advanced economies. There is a growing acceptance, in competitive business environments and project-based industries, that knowledge is a vital organisational and project resource that gives market leverage and contributes to organisational innovations and project success (Nonaka and Takeuchi, 1995; Egbu, 1999a, 2000). An organisation's capacity to innovate depends to a very considerable extent upon the knowledge and expertise possessed by its staff. It follows that, for leaders of construction organisations and projects, the processes by which knowledge is created or acquired, communicated, applied and utilised must be effectively managed. However, the point has been raised elsewhere (Egbu *et al.*, 2000) that our understanding of the important role of tacit knowledge in project innovations is limited, and so are the roles construction personnel play in knowledge management (KM) processes.

KM is emerging as a vital activity for organisations to preserve valuable knowledge and exploit the creativity of individuals that generates innovation. KM is important for a number of reasons. It is important because the rise of time-based competition as a marketing weapon requires organisations to learn quickly. It is important because of the globalisation of operations and because of the growth in number of mergers and take-overs where multiple organisations must share knowledge in a collaborative forum. In project-based industries the situation is even more complex. Project-based organisations are characterised by short-term working contracts and diverse working patterns. KM is important in this context because it brings together diverse knowledge sources from different sections of the demand and supply chains, achieving cross-functional integration. Understanding how organisations manage knowledge assets for improved innovations is important. There is, however, a paucity of empirical research on KM and its impact on innovation, especially in project-based industries such as construction.

KM is highly associated with innovation because of its ability to convert the tacit knowledge of people into explicit knowledge. This is grounded in the notion that unique tacit knowledge of individuals is of immense value to the organisation as a whole, and is the 'wellspring of innovation'. Given the close connection between knowledge possessed by personnel of the firm and the products and services obtainable from the firm, it is uncontroversial that a firm's ability to produce new products and other aspects of performance are inextricably linked to how it organises its human resources. It is tacit rather than explicit knowledge that will typically be of more value to innovation processes. Yet, tacit knowledge is knowledge that cannot be easily communicated, understood or used without the 'knowing subject'. The implication of the above discourse is that KM that focuses on creating network structures to transfer only explicit knowledge will be severely limited in terms of its contribution to innovation and organisational and project success.

Following this Introduction, the next section (8.2) discusses the intrinsic and complex relationships between KM and innovation and the role of building and maintaining capabilities to facilitate the process. Issues of strategy, process, structure, culture and technology and their impacts on KM are explored in Section 8.3. The implications for managers in managing knowledge for successful exploitation are discussed in Section 8.4 and conclusions are drawn in Section 8.5.

8.2 Knowledge management and innovations: building and maintaining capabilities

Innovation is a complex social process. No one best strategy exists or is suitable for managing innovations in every organisation. However, any meaningful innovation strategy should have unequivocal support from the top. Its objectives need to communicated and be accepted by the rank and file within the organisation. An innovation strategy needs to sit naturally within the overall strategy of the organisation. In addition, it is important that it is monitored and reviewed regularly.

Construction organisations need to determine their positions in terms of processes, services, products, technologies and markets. Since an organisation's innovation strategies are constrained by their current position, and by specific opportunities open to them in the future based on their competencies, construction organisations will need to determine their technological trajectories or paths. This will involve due cognisance of strategic alternatives available, their attractiveness and opportunities and threats, which lie ahead. The organisational processes that an organisation adopts in integrating the transfer of knowledge and information across functional and divisional boundaries (strategic learning) are essential and need to be consciously managed (Baden-Fuller, 1995). Since competitive advantage and financial success are bound up with industry dynamics, it is necessary to place strategic change in a competitive context and to identify what kinds of changes lead to strategic innovation, and when these changes result in benefits for the organisation.

Core capabilities and competencies are difficult to imitate and provide competitive advantage for organisations (Teece and Pisano, 1994; Tidd *et al.*, 1997). They are built through a knowledge-building process which is clustered around four learning styles: present problem solving, future experimenting and prototyping, internal implementing and integrating, and external importing of knowledge (Leonard-Barton, 1995). Resources and capabilities are keys to strategic advantage and organisations must build and maintain capabilities if they are to innovate. Similarly, an important capability is the expertise to manage internal and external organisational complementary resources. Through collaboration and by forming long-term relationships, construction organisations are able to learn from projects and transfer knowledge to organisational base and along supply chains.

Innovation might be thought of as a process of combining existing knowledge in new ways; often termed 'resource combination'. Resource combination depends on a cognitive process of 'generativity', which is the ability to form multipart representations from elemental 'canonical parts'. This cognitive integration or 'blending' is at the heart of the creation of novelty. Since generativity is in essence a combinatorial process, the more knowledge that we collectively accumulate, the more opportunities there are for the creation of innovative ideas.

Although the building of dynamic capabilities or core competencies is vital for organisational innovations, it is, however, important that core competencies do not turn into 'core rigidities', especially when established competencies become too dominant and important new competencies neglected or underestimated.

An organisation's competitive advantage can come from various sources, such as size and assets. In their study of four innovative construction organisations, however, Egbu *et al.* (1998) observed that the innovative organisations in their study were able to gain competitive advantage through their dynamic capabilities – by mobilising knowledge, experiences and technological skills. In the main, these were achieved through one or a combination of the following:

- focusing on a particular market niche
- novelty – offering something that no other organisation can
- complexity – difficulties associated with learning about their processes and technologies, which keeps entry barriers high
- stretching basic model of a product/process over an extended life and reducing overall cost
- continuous movement of the cost and performance frontiers
- integrating the person and the team around the product and service.

Construction organisations are, in the main, project-based organisations. Networking, communities of practice (CoPs), story-telling, coaching, mentoring and quality circles are important mechanisms for sharing and transferring tacit knowledge in project environments. These should be considered, encouraged and promoted more by construction personnel. CoPs are needed to encourage individuals to think of themselves as members of 'professional families' with a strong sense of reciprocity. The networking processes which can encourage sharing and the use of knowledge for project innovations are important. Leaders of construction organisations and projects should also espouse 'the law of increasing returns of knowledge' as a positive way of encouraging knowledge sharing. Shared knowledge stays with the giver while enriching the receiver.

Intuitive knowledge is managed by individuals being valued and not by being heavy-handed through project 'controlled processes'. It is folly to believe that any project organisation can make people have ideas and force them to reveal intuitive messages or share their knowledge in any sustained manner. An individual's intuitive knowledge cannot be manipulated in any meaningful way nor controlled without the individual being willing and privy to it. The process of trying to manipulate or control intuitive knowledge in fact creates its destruction. The issues of trust, respect and reciprocity are vital elements of a conducive environment for managing tacit knowledge. It is through these that individual members of the project can be motivated to share their experiences and exploit their creativity. Leaders of construction projects and organisations would need to recognise, provide incentives and reward knowledge performance and sharing behaviour patterns. Leaders should also take action on poor knowledge performance. The regular communication of the benefits of KM is important in sustaining the co-operation of project team members. A variety of ways exist for doing this, including regular meetings, project summaries, project memos and through project Groupware/extranet facilities where they exist. Every project strategy for KM should consider the training, recruitment and selection of project team members (e.g. subcontractors and suppliers). It should also pay due cognisance to the team members' competencies, requisite knowledge and their willingness and effectiveness in sharing knowledge for the benefit of organisations and projects. In addition, the 'absorptive capacity' of the parties involved in the knowledge-sharing processes is vital.

8.3 Knowledge management and improved innovations: issues of strategy, process, structure, culture and technology

Knowledge management impacts upon organisational and project innovations in many complex ways through a host of interrelated factors. An understanding of these factors and their contribution to innovations is important for the competitive advantage of project-based organisations.

A good internal organisational structure, expressed through the strategies, processes and culture of an organisation, is one that is flexible but supportive of the ideas propounded by employees. The organisational structure should respond just as effectively to external pressures. Hierarchical structures become deficient in turbulent environments. In contrast, structures determined by core competencies can adapt to chaotic external pressures more easily. Such competencies should be flexible to meet new customer demands or exceed expectations. Excessive bureaucracy can stifle innovation because of, for example, the amount of time it takes to approve every idea. While in small organisations this may require minimal bureaucracy, in larger, more complex organisations the process is incessantly cumbersome. Organisational structures need to sustain equilibrium between creativity and formal systems. Bureaucracy can inhibit spontaneity and experimentation and thus threaten the innovation process. However, bureaucratic structures may also assist the rapid and continuous transformation of ideas into superior products.

While the centralisation of an organisation's structure for decision making can create a definite medium of control, a more informal and flexible structure is desirable for knowledge generation. Flexible structures encourage better internal communications and a more change-friendly climate where ideas and knowledge are shared freely.

As aforementioned, the tacit knowledge of the individual is an essential component of organisational success. However, such knowledge is often guarded by those who are reluctant to transfer this 'power' from an individual level to the organisational level. Therefore, the employee must be sufficiently motivated to share knowledge, through incentives. The organisational structure should play a part in the encouragement of knowledge sharing. It is contended that motivation is a key facilitator of loyalty and trust amongst employees and eventually fosters continuous learning.

Every manager has a vision of the organisation he/she works for. The importance of expressing this vision to the rest of the organisation is paramount. There is need for a long-term vision to be incorporated into the corporate strategy of the company. This is only achievable if the context of the organisation is fully understood. Sullivan (1999) identifies three key areas to be understood. First, what are the real features of the business, i.e. the core competencies? Second, what is the external context such as the socio-political and economic forces of change and their particular impact on the company? Third, what is the internal context, e.g. the strategy, culture, performance, strengths and weaknesses of the company? In sum, Sullivan asserts the need for effective management of a company's capabilities, e.g. management of portfolio (intellectual property and intellectual assets), competitive assessment, human capital management. This has the potential for improving organisational innovations, leading to competitive advantage and market leverage.

Project-based industries, especially the construction industry, are under growing pressure to compete in new ways. Strategic planning and the need for growth are seen to require organisations to develop firm-specific patterns of behaviour, i.e. difficult to imitate combinations of organisational, functional and technological skills. These unique combinations create competencies and capabilities and take place as the organisation's intangible knowledge is being applied in its business behaviour, especially in its value-adding business process. Competitive advantage stems from the firm-specific configuration of its intangible knowledge. The development of core competencies or capabilities creates an environment of strategic thinking, in which knowledge and ideas are key.

Much of the literature on innovation focuses on the need to establish the right kind of organisational culture. It would be a mistake to underestimate the importance of cultural factors in the adoption of KM and organisational learning. A climate favourable to innovation must be achieved by committing resources, allowing autonomy, tolerating failure and providing opportunities for promotion and other incentives. Thus, an organisation must be flexible enough to facilitate the innovation process. In order to establish a knowledge-based organisation there needs to be a supportive organisational culture. The cultivation of a 'learning organisation' is an essential requirement for knowledge managers. If an organisation develops a learning culture, there is scope for both formal and informal channels of 'dialectic thinking', where individuals develop their individual capabilities through positive experimentation. Further theories about organisational culture favour the evolution of a 'community of practice' where social interaction of employees cultivates a knowledge-sharing culture based on shared interests, thus encouraging idea generation and innovation.

In all organisations, the politics of knowledge sharing is an issue. Employees and employers from diverse backgrounds often come into conflict over important decisions. Manipulating these tensions to achieve 'creative abrasion' is a strategy to maximise innovation. However, it is a challenging task that involves disciplined management. Leadership is an inherent part of organisational culture, but also extends into areas of strategy and structure. Leadership is an organisational responsibility. The value of institutional leadership must be its ability to create the structures, strategies and systems that facilitate innovation and organisational learning. Organisations should build commitment and excitement, collective energy and empowerment. There is need for a managerial commitment to the long-term strategic vision of an organisation and the motivation to achieve the goals set out. Moreover, empowering employees to generate and share knowledge is the task of management. For example, implementation of rewards and punishment schemes are stimuli for successful KM. Motivating employees to share the knowledge they have involves good people management, where trust is itself an incentive. The establishment of a psychological contract between employer and employee, for example, is a constructive approach to developing a knowledge-sharing culture (Scarbrough *et al.*, 1999).

KM is about mobilising the intangible assets of an organisation, which are of greater significance in the context of organisational change than its tangible assets, such as information and communications technologies (ICT).

While information technology is an important tool for a successful organisation, it is often too heavily relied upon as a guarantee of successful business. Such tools as the internet are merely enablers. The true asset of an organisation is the brainpower of its workforce. It is the intellectual capital of an organisation that is the key to success. Thus, KM is not just about databases or information repositories. In computer systems the weakest link has always been between the machine and humans because this bridge spans a space that begins with the physical and ends with the cognitive. Notwithstanding this, the important role of ICT in knowledge (especially for explicit knowledge) capture and retrieval and their implications for innovations in the construction industry have been well documented (Egbu, 2000; Egbu and Botterill, 2002).

8.4 Managing knowledge for exploiting innovations: implications for managers

If KM is to have any real impact on the way construction organisations do business, then it has got to be about making radical changes in the way organisations utilise knowledge. Knowledge has to be 'made productive'. Managers have critical roles to play in making knowledge productive, in knowledge development and in the exploitation of knowledge for innovative performances. Deepening the understanding and analysis of a manager's interest in knowledge is vital in order to understand how KM can contribute to improved strategic formulation. The following are key issues for consideration by management:

- Make KM an integral part of strategic decisions on profitability and competitiveness of the organisation. In this regard, establish at all levels of the organisation a strategic intent of knowledge acquisition, creation, accumulation, protection and exploitation of knowledge. The linkages between strategic management and human value need to examine carefully the role of a KM orientation as an effort to support adequately successful strategies.

- Management should not only be interested in knowledge development but proactively support it. As part of knowledge development, it is important that knowledge workers (organisation and project staff and team members) are included in a dynamic KM process, which demands the support of motivation, creativity and the ability to improve a comprehensive vision of the relationship between the organisation, project and its environment.

- Determine appropriate mechanisms for the effective capture, transfer and leveraging of knowledge. Communication infrastructure needs to be established within and between the different departments and strategic business units. This should support and enhance the transfer of ideas, at the same time not limiting the potential for creativity and the questioning of actual activities that are needed to understand the challenges in the wider environment, and may be a source of new solutions to problems.

- Create an appropriate culture for effective KM. There is a need to encourage the workers' autonomy, so that they may express and share the knowledge they possess in a 'free environment'. This should also aim for the assimilation of external knowledge with internal thoughts, ideas and experiences, avoiding the effects of not-invented-here syndrome.

- Determine methods for measuring the extent of KM effectiveness. Audit the knowledge present at, or accessible to, the organisation and manage adequately the inventory of 'knowledge repositories'.

8.5 Conclusions

The important roles of KM have been considered against the new differentiators of success in competitive environments, namely accelerated innovation and dynamic core capabilities. When managed effectively, innovation creates possibilities for competitive advantage.

KM should be seen as a long-term strategic concern. The effective management of knowledge assets involves a holistic approach to a host of factors, including human, technological, economic, social, environmental and legal issues associated with the creation, sharing, transferring, storing and exploitation of explicit and implicit knowledge dimensions. Also a host of factors combine in different ways to produce successful organisational innovations. These include having the ability to manage organisational knowledge (tacit and explicit) and build knowledge-enhancing approaches, systems and technology, and integrate the person and the team around the product. If the construction industry is to build core competencies, maintain capability and benefit from innovation, it has to change from an adversarial and blame culture to a sharing culture.

Management has a very important role to play in the exploitation of KM for improved innovations. There also needs to be more targeted education and training programmes for practitioners, which should reflect the nature of innovation and KM dimensions as very complex social processes.

References

Baden-Fuller, C. (1995) Strategic innovation, corporate entrepreneurship and matching outside-in to inside-out approaches to strategy research. *British Journal of Management*, **6**, Special Issue, December, S3–S16.

Egbu, C.O. (1999a) Mechanisms for exploiting construction innovations to gain competitive advantage. In: *Proceedings of the 15th Annual Conference of the Association of Researchers in Construction Management (ARCOM)*, Liverpool John Moores University, 15–17 Sept, vol. 1, pp. 115–23.

Egbu, C.O. (1999b) The role of knowledge management and innovation in improving construction competitiveness. *Building Technology and Management Journal*, **25**, 1–10.

Egbu, C.O. (2001a) Managing innovation in construction organisations: an examination of critical success factors. In: *Perspectives on Innovation in Architecture, Engineering and Construction* (C.J. Anumba, C. Egbu and A. Thorpe, eds), pp. 1–9. Centre for Innovative Construction Engineering, Loughborough University.

Egbu, C.O. (2000) The role of information technology in strategic knowledge management and its potential in construction industry. In: *Proceedings of a UK National Conference on Objects and Integration for Architecture, Engineering and Construction*, BRE, Watford, 13–14 March, pp. 106–114.

Egbu, C. and Botterill, K. (2001) Knowledge management and intellectual capital: benefits for project-based industries. In: *Proceedings of the RICS COBRA Conference*, 3–5 Sept, Glasgow Caledonian University, vol. 2, pp. 414–22.

Egbu, C. and Botterill, K. (2002) Information technologies for knowledge management: their usage and effectiveness. *Journal of Information Technology in Construction*, ITcon, **7**, Special Issue ICT for Knowledge Management in Construction, pp. 125–37 (available at http://www.itcon.org/2002/8).

Egbu, C.O., Henry, J., Quintas, P., Schumacher, T.R. and Young, B.A. (1998) Managing organisational innovations in construction. In: *Proceedings of the Association of Researchers in Construction Management (ARCOM) Conference*, 9–11 Sept, Reading, vol. 2, pp. 605–614.

Egbu, C.O., Gorse, C. and Sturges, J. (2000) Communication of knowledge for innovation within projects and across organisational boundaries. In: *Proceedings of the 15th International Project Management World Congress*, Royal Lancaster Hotel, London, 22–25 May.

Gann, D.M. (2000) *Building Innovation: Complex Constructs in a Changing World*. Thomas Telford, London.

Leonard-Barton, D. (1995) *Wellspring of Knowledge: Building and Sustaining the Sources of Innovation*. Harvard Business School Press, Boston, Massachusetts.

Nonaka, I. and Takeuchi, H. (1995) *The Knowledge Creating Company: How Japanese Companies Create the Dynamics of Innovation*. Oxford University Press, Oxford.

Scarbrough, H., Swan, J. and Preston, J. (1999) *Knowledge Management: A Literature Review. Issues in People Management*. Institute of Personnel Development, London.

Slaughter, E.S. (1993) Builders as sources of construction innovation. *Journal of Construction Engineering and Management*, **119**(3), 532–49.

Sullivan, P.H. (1999) Profiting from intellectual capital. *Journal of Knowledge Management*, **3**(2), 132–42.

Tatum, C.B. (1987) Process of innovation in construction firms. *Journal of Construction Engineering Management*, **113**(4), 648–63.

Teece, D. and Pisano, G. (1994) The dynamic capabilities of firms: an introduction. *Industrial and Corporate Change*, **3**, 537–56.

Tidd, J., Bessant, J. and Pavitt, K. (1997) *Managing Innovation: Integrating Technological, Market and Organisational Change*. John Wiley and Sons, Chichester.

Wolfe, R.A. (1994) Organizational innovation: review, critique and suggested research directions. *Journal of Management Studies*, **31**(3), 405–431.

9 Performance Measurement in Knowledge Management

Herbert S. Robinson, Patricia M. Carrillo,
Chimay J. Anumba and Ahmed M. Al-Ghassani

9.1 Introduction

There is a wealth of literature advocating the benefits of managing knowledge in organisations. Those with responsibilities for implementing knowledge management (KM) strategies such as chief knowledge officers and knowledge managers are increasingly being challenged to make a business case for KM given the competing needs for organisational resources. The benefits of KM include enabling organisations to learn from the corporate memory, share best practice knowledge and identify competencies to become a forward-thinking and learning organisation. Demarest (1997) noted that 'firms without knowledge management systems will be effectively unable to achieve the re-use levels required by the business model implicit in the markets they enter, and will lose market share to those firms who do practise knowledge management'. Leading management consulting firms have invested substantial effort trying to develop KM systems to package, use and reuse knowledge. Many other organisations including construction firms are gradually embracing the concept and practices of KM.

Construction as a knowledge-based industry has been addressed in Chapter 2. Previous chapters have also discussed the nature of knowledge, and importance of KM, the need for developing a KM strategy and the processes and tools required for its implementation. However, a key issue for organisations is how to evaluate the performance of KM projects, programmes or initiatives and associated knowledge assets. Thomas Stewart, editor of the influential *Harvard Business Review,* argued that 'Knowledge may be intangible but that doesn't mean it can't be measured'. Evaluating or measuring KM is also a major challenge, which although feasible is very difficult in practice. Performance measurement is desirable given the increased investment in KM activities and the need to demonstrate its practical value. However, there is no universally accepted method, as the development of performance metrics for knowledge and KM is a rapidly evolving area.

This chapter focuses on performance measurement in a KM context. Following the Introduction, Section 9.2 explains the reasons for a performance-based approach in KM. Section 9.3 discusses two distinct but interrelated aspects of knowledge that need to be measured: the first relating to knowledge assets (*stocks*), and the second to knowledge management projects, programmes or initiatives (*flow*) aimed at improving or increasing the value of organisational knowledge stocks, which in turn influences business performance. Section 9.4 explores the different approaches for measuring or evaluating KM and associated knowledge assets. The underlying principles of various measurement techniques are discussed, and a selection of practical tools to evaluate the performance of KM and knowledge assets are outlined and compared (Section 9.5). The chapter concludes (Section 9.6) that measurement is central to evaluating knowledge assets and KM activities, and different measurement techniques and types of measures could be used in a complementary way to evaluate performance.

9.2 Why measure the performance of knowledge management and knowledge assets?

The strongest argument for measuring the performance of knowledge assets and KM is to demonstrate its business benefits so that the resources and support necessary for a successful implementation can be provided. Previous chapters have shown that construction organisations could benefit significantly from implementing KM during the entire project cycle, from planning, design and construction to facilities management. For example, a leading engineering consulting organisation highlighted that feedback from their legal department shows that the single largest cause of loss of money within the firm was a failure to agree the appropriate contract terms up front (Sheehan, 2000). The knowledge manager explained that a KM system such as the collation of a legal intranet page pushed to the desktop at appropriate times in projects is an increasingly effective solution to this problem. KM initiatives could also facilitate the smooth running of construction contracts. A leading firm of solicitors developed a stand-alone hypertext system based on the standard form of international construction and engineering contract (FIDIC), with context-sensitive expert commentary provided by specialist lawyers (Terrett, 1998). It is argued that this system could lead to significant savings in legal costs often associated with construction projects. Table 9.1 shows some cost savings due to KM programmes that some leading organisations have achieved.

The old maxim 'you cannot manage what you cannot measure' also applies to knowledge. Measurement is therefore an integral part of KM as it provides information and feedback on the relationship between knowledge assets (stocks) and KM projects or initiatives (flows). The role of KM is to improve the value of knowledge stocks, and to indirectly improve business performance or the market value of an organisation. Knowledge assets represent a significant proportion of the market value of some organisations. There is now an increasing realisation of the importance of knowledge, sometimes referred to as intangible assets or intellectual capital. Skandia, a

Table 9.1   Examples of cost savings from KM programmes

- Texas Instruments saved itself the $500 million cost of building a new silicon wafer fabrication plant by disseminating best internal working practices to improve productivity in existing plants.

- Skandia AFS reduced the time taken to open an office in a new country from seven years to seven months by identifying a standard set of techniques and tools that could be implemented in any new office.

- Dow Chemical has generated $125 million in new revenues from patents and expects to save in excess of $50 million in tax obligations and other costs over the next ten years by understanding the value of its patent portfolio and actively managing these intellectual assets.

- Chevron Oil made savings of $150 million per year in energy and fuel expenses by proactive knowledge sharing of its in-house skills in energy use management.

large Swedish insurance and financial services company, was the first organisation to introduce an intellectual capital report to persuade investors of the value of an organisation's knowledge (Edvinsson, 1997). The value creation concept showing the relationship between tangible and intangible assets (knowledge) of an organisation is illustrated by Skandia's tree metaphor in Figure 9.1.

'Hard physical' assets	**Tangible assets**	**Debt Equity**	Financial capital
'Soft knowledge' assets	**Goodwill Technology Competence**	**Intellectual capital**	Non-financial capital

Figure 9.1   Skandia's tree metaphor (adapted from Edvinsson, 1997)

Like the roots of a tree, knowledge is the hidden asset of organisations, which has to be nurtured for long-term corporate sustainability. Intellectual capital includes assets such as brands, customer relationships, patents, trademarks and, of course, knowledge. Tangible assets such as buildings, plant and equipment remain essential for the production of goods and services, and so is the capital required (debt equity) for financing business operations. These measures are normally incorporated in conventional balance sheets and accounting systems. So a manager is able to provide information very quickly on how much a company has in the bank, the value of its land, plant and buildings, working capital and inventories. However, information on knowledge assets, considered to be the greatest source of wealth, the driver for innovation and competitive advantage, is often not available. This is because intellectual capital, technology competence and goodwill are not reflected in traditional measurement frameworks such as the balance sheet and profit and loss accounts. There is, at least, a recognition of goodwill, but only when a company is sold or bought for more than its book value because of accounting conventions or rules. However, reliance on tangible assets alone as key performance measures can be misleading at best, resulting in short-term benefits. This is often at the expense of understanding the key measures of knowledge (intellectual capital or intangible assets) that influence financial performance.

While most if not all knowledge-based organisations recognise the value of knowledge, and the need to bring it to the centre stage, very limited progress has been made in measuring soft knowledge to supplement traditional financial information. There is growing evidence that non-financial measures relating to intangibles or knowledge are becoming important to organisations, investors, shareholders, employees and other stakeholders. A survey of North American, European and Asian businesses found that 89% of the organisations sampled (regardless of size) agreed with the statement that 'measuring intellectual capital will be critical to the organisation's ability to achieve business success' (Bontis, 2000).

Thomas Stewart, editor of the influential *Harvard Business Review*, in his book entitled *Intellectual Capital: The New Wealth of Organisations* (Stewart, 1997), argued that 'it would be a mistake to mingle measures of intellectual capital with financial data, it would be a greater one not to use them at all'. While financial measures will continue to be a crucial aspect of corporate performance there is evidence that non-financial measures relating to intangibles and knowledge assets are becoming increasingly important for corporate sustainability. This is due, in part, to the growing interest and demand from investors, clients and other stakeholders for changes in corporate reporting following recent high-profile business failures. Samuel DiPiazza Jr, Chief Executive Officer of Pricewaterhouse Coopers, and Robert Eccles, former professor at Harvard Business School, in their book entitled *Building Public Trust: The Future of Corporate Reporting*, argued that organisations should provide a broader range of information than financial reporting regulations require (DiPiazza and Eccles, 2002).

9.3 Types of performance measures

Measurement facilitates benchmarking and the identification of best practices in KM. Two distinct aspects of KM need to be measured. The first relates to knowledge assets (*stocks*), and the second to KM projects, programmes or initiatives (*flow*) aimed at improving or increasing the value of knowledge assets (stocks). The knowledge stocks are, for example, the talents of people employed, the efficiency of the processes used, the nature of products and customer relationships. Bontis *et al.* (1999) argued that 'there is a positive relationship between the stocks of learning (knowledge) at all levels in an organisation and its business performance'. Investment in developing organisational competencies (knowledge stocks) may, therefore, be wasted if the flow of learning is obstructed, i.e. if there is no KM or it is not properly implemented. KM facilitates the flow of learning between knowledge stocks – individuals, groups, business processes, customer relationships and products – and thereby increases the value of an organisation's knowledge assets.

9.3.1   Measures for knowledge assets

Measures for knowledge assets relate to what organisations know, and what they need to know or learn to improve performance. This is embedded in an organisation's processes, products and its people. The objective is to identify the most significant measures that correlate with particular business objectives. For example, if the strategic objective of a construction organisation is to expand, measures relating to procurement, bid/win ratio and workload could be more significant. If, on the other hand, the strategic objective is to maximise stakeholders' involvement, then measures such as staff turnover, customer satisfaction and public perception could be given more significance.

Measures for knowledge assets/stocks or intellectual capital focus on several main components – human, structural and customer capital. Human capital is the knowledge in people's heads, acquired mainly through education, training and experience. Structural capital is the knowledge embedded in business processes, so-called non-human storehouses, including organisational manuals, procedures and databases. Customer capital refers to knowledge about products, marketing channels and customer relationships.

9.3.2   Measures for knowledge management

Measures for knowledge management focus on the expected outputs of KM interventions or initiatives relative to its inputs. A *full* evaluation involves a comparison of both the inputs and the outputs of KM interventions. For example, to improve the bid/win ratio and workload in a construction organisation, it may be necessary to improve knowledge sharing. Alternative initiatives could include: (1) setting up a post-tender forum with clients for sharing information with the estimating team, (2) creating a community of practice for estimating involving the entire supply chain, or (3) developing a business-intelligent system to support the estimating process. A variety of performance measures could be adopted to evaluate KM alternatives. It is easier to compare performance using cost-benefit analysis where measures of inputs and outputs are included or transformed to the same monetary units. This may also facilitate a comparison between business units of the same organisation or different organisation. However, where units of inputs and outputs are different and difficult to transform to the same units, other performance measures such as cost-utility and cost-effectiveness measures could be adopted.

9.4   Measurement approaches

Case study investigations with leading construction organisations show that KM is increasingly recognised as important. However, there are two main problems: first, the links between KM and business performance are not clearly understood; second, there is often a lack of appropriate tools to track the impact of KM and knowledge assets. It is, therefore, not surprising that there are considerable difficulties associated with demonstrating the benefits of KM. Many organisations have recognised this, but appropriate methods are not always put in place to evaluate the performance of knowledge assets and KM projects, programmes or initiatives.

There is currently no universal standard for measuring or evaluating knowledge assets and/or KM programmes. Choosing an appropriate tool for measurement is crucial in assessing the effectiveness and efficiency of an organisation's knowledge assets and KM programmes. A variety of performance measurement approaches have emerged that can be grouped into (1) metrics, (2) economic and (3) market value approaches. The underlying principles and the limitations of these methods are discussed below.

9.4.1   Metrics approach

Metrics are input or output indicators used to monitor the performance of knowledge assets or KM programmes. On the input side, the indicators reflect enablers or actions required to implement or achieve business objectives. Examples of input indicators are number of training days per employee, proportion of staff with professional qualifications (i.e. chartered) or with over two years' experience, and senior managers with experience on major projects. The output indicators measure the performance or the result of those actions, such as the number of defects after project completion, complaints from clients, and cost and time overruns. Metrics can be single or composite, i.e. an aggregate of individual indicators into a single index such as the Intellectual Capital (IC) index. This approach is based on the assumption that there is a relationship or correlation between the indicators and business performance. As Stewart (1997) noted, 'if you cannot demonstrate the link between increased customer satisfaction and improved financial results, you are not measuring customer satisfaction correctly'. By measuring certain attributes or components of a business such as employees, processes, customers, etc., valuable organisational knowledge could be

Table 9.2   Examples of metrics

Metrics type	Metrics
Human	Employee satisfaction (e.g. absenteeism, job security) Training and experience (e.g. education, project managers on major assignments) Knowledge networks (e.g. communities of practice) Knowledge worker turnover rate
Structural	Innovation (e.g. research collaboration, patents, trademarks) IT infrastructure (e.g. volume of knowledge content, usage) Bidding process (e.g. bid/win ratio) Construction process (e.g. defects, waste, pollution) Safety procedures (e.g. accidents)
Customer	Customer satisfaction Loyal customers (e.g. repeat business) Number of customers gained versus lost Business intelligence (knowledge about competitors, customers and markets)

provided to improve business performance. For example, by measuring participating rates in communities of practices and defect rates, customer complaints could trigger early warning signals for corrective actions to be taken. Metrics incorporate both financial and non-financial measures influencing business performance, and are generally grouped into three categories: customer, structural and human capital. Examples of specific metrics are given in Table 9.2.

- *Customer capital*: focus is on measures associated with products, such as customer satisfaction, attracting and retaining customers, the deployment of new products and services, reduced cycle time (of product or services to market), sales, revenue growth and market share, and the development of detailed customer knowledge.
- *Structural capital*: focus is on measures for processes associated with improved resource allocation, cost reduction, increased productivity by leveraging best practices across the organisation, innovative processes, business processes or services.
- *Human capital (people)*: focus is on measures associated with people and people sharing competencies, retained expertise, new recruits, retirement, job changes and transfers, commercialisation of ideas as new products and number of creative solutions in networks.

A number of application tools are underpinned by the metrics approach. Those include the Skandia Navigator (Edvinsson, 1997) and the Intangible Assets Monitor (Sveiby, 1997a), and business performance measurement models such as the Balanced Scorecard by Kaplan and Norton (1996) and the Excellence Model (EFQM, 1999).

Limitations

A number of problems are associated with the metrics approach. First, it is often difficult to aggregate or combine different metrics into a single numeric measure to correlate with business performance. Second, comparison between business units or organisations can sometimes be difficult, if not meaningless, without a standard metric definition for each measure. For example, the concept of key performance indicators (KPIs) in construction is an attempt to standardise certain measures, but there are many other measures used by different organisations without a standard definition. Third, and perhaps most significantly, metrics do not always provide adequate information about performance to enable continuous improvement initiatives to be undertaken. For example, the number of skips may inform the finance/accounting department about the level of waste in monetary terms, but such information is of limited value to the environmental department. 'Designing out' waste is a major issue for many construction firms, but it is important to understand the nature of waste – composition by type of materials, void space, causes of waste, etc. – in order to develop the most appropriate strategy for waste reduction.

9.4.2   Economic approach

While the metrics approach is based on the implicit assumption that improvement in metric scores could result in improved business performance, economic approaches are more explicit and attempt to calculate the actual contributions or net benefits. Table 9.3 provides examples of the benefits associated with some performance metrics.

However, this approach goes beyond anticipating the benefits and recognises that the costs associated with KM are crucial, and the objective is to assess whether the benefits exceed the costs.

The weighted sum of inputs in relation to the weighted sum of outputs, expressed in monetary values, utility values or other units of measurement, is an indication of the measure of performance. A range of KM tools, i.e. techniques and technologies (discussed in Chapter 6), are used to support the implementation of KM strategies and initiatives. The costs and benefits of using these tools vary enormously. For example, introducing a document management system will involve information technology inputs (hardware and software costs) as the most obvious, but also human costs are associated with setting up and maintaining the system. The outputs or benefits are the direct cost savings and could include consumables such as printing, papers, travel costs, delivery costs, as documents are available on-line for discussions, revisions and approval. The costs and/or benefits could be one-off, on-going, up-front or back-end. The method is therefore useful in determining the likely returns on investment (ROI), internal rate of return (IRR), net present value (NPV) or payback period to help managers determine the value of KM.

Table 9.3   Examples of metrics and associated benefits

Performance measures	Metric definition	Expected benefit
Staff retention/ staff turnover	Percentage of staff retained or leaving	Reduction or increase in staff recruitment costs
Safety	Number of reportable accidents per 100 000	Reduction in accident costs
Productivity	Output/turnover per employee; value added per employee	Increase/decrease in turnover/output
Absenteeism	Percentage of days absent per employee	Reduction in the cost of absenteeism
Complaints/ compliments	Number of complaints/ compliments from customers	Potential loss/gain of business opportunities
Defects	Number of major defects	Reduction in cost of defects
Repeat business	Value of repeat business as percentage of turnover	Increase in the value of repeat business
Bidding – bid/win ratio	Number of bids won out of total submission/decisions known	Reduction in the cost of tendering
Waste	Quantity of waste/number of skips	Reduction in landfill charges, fuel-related cost
Noise pollution	Numbers of complaints/notices issued/fines	Reduction in sanctions/fines

Economic approaches could also involve the valuation of specific knowledge assets or components; for example, quantifying the economic value of people to an organisation where human capital comprises a significant proportion of organisational value, or other intangibles.

There are a number of application tools incorporating various techniques for quantification of the value of KM. Examples include the IMPaKT Assessor (Robinson *et al.*, 2002; Carrillo *et al.*, 2003), the Benefits Tree and the Inclusive Value Methodology (Skyrme, 1998).

Limitations

This approach has some shortcomings with regard to the methodology and interpretation of the outcome. While assessing the cost of KM may be relatively straightforward, the quantification of benefits is sometimes problematic, cumbersome and difficult. For example, in determining labour savings and savings on other expenditures such as printing, papers, telephones and travel costs, productivity increases may involve assumptions, some more plausible than others. Such methods, although useful in determining the value of KM, tend to rely extensively on 'guestimates' or data that are sketchy or sometimes not available. The approach is often imprecise and therefore requires caution in the interpretation of the outcome, as the exact contribution of KM practices to the improvement of a particular aspect of business, with respect to other factors, is not always certain.

9.4.3 Market value approach

Metrics and economic approaches are 'micro' in focus or disaggregated, i.e. they concentrate on particular aspects of KM or dimensions of knowledge assets. Market value approaches, on the other hand, are 'macro' in nature, focusing on the whole organisation, the aggregate of knowledge assets or market factors. The market value approach is based on the principle that the value of a company comes from its hard financial capital (physical and monetary assets) and soft knowledge or intellectual capital. Knowledge or intellectual capital should therefore explain the difference between the value assigned to an organisation by a buyer or the stock market in relation to its book market value.

Existing accounting/measurement frameworks focus on hard tangible assets and their costs, i.e. the production side, rather than soft intangible assets of knowledge and the creation of value. Research and development (R&D) and intellectual assets now form a significant component of the costs in modern knowledge organisations, compared to materials and labour in traditional industrial organisations. Existing measurement frameworks are therefore considered grossly inadequate if not misleading.

KM researchers and practitioners believe that the growing discrepancy between market value and book value is largely attributed to intellectual capital (where the market value exceeds book value) or intellectual liabilities (where the book value exceeds market value). There is evidence of the market values significantly exceeding book values in certain business sectors that are knowledge-intensive, such as management consulting, biotechnology, pharmaceuticals, information technology and software development services. For example, in 1995, IBM paid US$3.5 billion for Lotus, which represented seven times its book value (Jordan and Jones, 1997). This is a reflection that the physical, visible or hard assets contribute significantly less than the hidden, soft assets of knowledge. As Stewart (1997) puts it, 'you don't buy Microsoft because of its software factories; the company doesn't own any. You buy its ability to write code, set standards for personal-computing software, exploit the value of its name and forge alliances with other companies'.

Limitations

A fundamental criticism of the market value approach relates to the vagaries and volatility of the stock market, often responding to factors outside the control of companies and their management. The approach is generally highly aggregated and, therefore, limited in value, in terms of the details of the information provided for understanding and guiding knowledge assets and KM strategies in individual company situations and circumstances.

9.5   Application tools

An overview of some of the common measurement application tools is provided below, and a comparison of selected tools is presented later in Table 9.5. More detailed information about specific measurement tools can be found in the references provided.

9.5.1   Business performance measurement models

Business performance measurement models are approaches incorporating a variety of financial and non-financial measures. Examples include the Balanced Scorecard developed in the Harvard Business School by Kaplan and Norton (1996) and the Excellence Model developed by the European Foundation for Quality Management (EFQM, 1999).

The Balanced Scorecard is designed to focus managers' attention on those factors that help the business strategy; it adds, alongside financial measures, measures for customers, internal processes and innovation. The Excellence Model encourages organisations to adopt a forward-looking approach by focusing on a broad range of metrics reflecting the wider business environment. It incorporates measures for leadership, people, processes, policy and strategy, partnership and resources that influence people, customers, society and key performance results.

Both the Balanced Scorecard and the Excellence Model are performance measurement systems that can be adopted for measuring KM as they have learning components linked to organisational performance. In contrast to traditional accounting practices, these models complement financial measures by including measures of intangible assets reflecting the components of intellectual capital.

9.5.2   Skandia Navigator

This method, developed by Skandia, a large Swedish financial services company, focuses on five areas: (1) financial, (2) customer, (3) process, (4) renewal and development, and (5) human aspects, with metrics for each area (Edvinsson, 1997).

The leading advocate *Leif Edvinsson* asserts that the value of intangible assets is significant and contributes to the widening gap between a

Table 9.4   Sample measures for the Intangible Assets Monitor

	External structure	Internal structure	Competence of people
Growth and renewal	• Growth in market share • Satisfied customers	• Time devoted to R&D • Investment in IT	• Growth in average professional experience • Competence turnover
Efficiency	• Profit per customer • Sales per employee	• Proportion of support staff • Sales per support staff	• Change in value added per employee
Stability	• Proportion of big customers • Frequency of repeat orders	• Age of organisation • Support staff turnover	• Average age of employees • Seniority

company's book value or corporate balance sheet and investors' assessment of the market value. The approach therefore measures the hidden dynamic factors of human, customer and structural capital that underpin the visible aspects of the company's buildings and products, to reflect more accurately the value of the company.

9.5.3   Intangible Assets Monitor

The intangible Assets Monitor shown in Table 9.4 was pioneered by Karl Erik Sveiby (Sveiby, 1997a,b) as a framework based on three families of intangible assets: (1) external structure (brands, customer and supplier relations); (2) internal structure (the organisation: legal structure, manual systems, attitudes, R&D, software); and (3) individual competence (education, experience). Three sets of measurement indicators are also provided for each of the three families of intangible assets: growth and renewal, efficiency and stability. The approach is based on the principle of replacing the traditional accounting framework with a new measurement framework underpinned by a knowledge perspective to provide a complete indication of financial success and shareholder value.

9.5.4   Human Resource Accounting

The Human Resource Accounting (HRA) approach is based on the principle that human assets, which reflect the skills, know-how and experience embedded in the workforce, are a proxy for capital, and can be calculated and reflected in the balance sheet by capitalising salary expenditures. The approach involves quantifying the economic value of people in an organisation using one of three types of HRA measurement models: cost models, human resource value models and monetary emphasis models (Bontis *et al.*, 1999). Cost models consider the replacement or opportunity cost of human assets. Human resource value models combine non-monetary with monetary economic value models. Monetary emphasis models calculate discounted estimates of future earnings or wages.

9.5.5 IMPaKT Assessor

Given that the purpose of KM is to improve business performance, the IMPaKT (Improving Management Performance through Knowledge Transformation) approach logically ties in KM to an organisation's business strategy. It is a three-stage framework developed to facilitate an understanding of KM implications of business problems, and to assess the impact of KM on business performance. The approach, underpinned by a cause-and-effect map, links KM initiatives to the strategic business objectives they are related to and performance metrics. It also provides a road map for selecting the most appropriate evaluation technique for the quantification of the value of KM. IMPaKT was developed at Loughborough University in collaboration with leading construction organisations (Robinson *et al.*, 2002; Carrillo *et al.*, 2003).

9.5.6 KM Benefits Tree

The Benefits Tree shows the interdependencies between different types of knowledge benefits, intermediate and organisational benefits (Skyrme Associates, 2003). It provides a systematic approach to help users understand the factors that influence the outcome of KM so that investment in KM can be justified.

9.5.7 Degussa–Huls approach

This is similar to the Benefits Tree approach but provides a cause-and-effect diagram to analyse and measure the value of KM activities in quantitative and qualitative terms (Bunz and Kirch-Verfuss, 2001). The cause-and-effect network consists of three components. The first component relates to the priorities of the KM initiatives. The second component consists of possible transfer effects on six dimensions of KM: people, management, processes, technology, innovation, customers and market. The third part reflects a variety of business results, with earnings before interest and tax (EBIT) as the key financial measure. Using the tool involves linking the three components, adding impact factors in each link and summing up the individual factors to arrive at an overall impact factor for each initiative at the third level – the business and financial level.

9.5.8   Department of Navy, USA, (DON) Approach

This is a framework for measuring the value of KM initiatives and to help managers identify and apply appropriate metrics for assessing performance (Department of Navy, 2001). Three types of measures are used to measure the performance of KM initiatives from different perspectives: outcome metrics, output metrics and system metrics.

9.5.9   Inclusive Value Methodology (IVM)

This approach is a multi-dimensional process that incorporates formal accounting practices for financial measures of value but is extended, based on the same principles, to deal with non-financial performance measures or intangibles (Skyrme, 1998). It was developed by Professor Philip McPherson, emeritus professor at City University.

9.5.10   Market-to-book-value ratio

The market value is what a company is worth as a whole and the book value is what is left of a company after taking account of the debt owed. This method is based on the simple assumption that what is left in a company after accounting for hard fixed assets are the soft intangible assets or knowledge capital.

9.5.11   Tobin's q ratio

This approach is based on the work of Nobel prize winner and economist James Tobin. It compares the market value of an asset with its replacement cost (book value). The method was developed to help predict corporate investment decisions independent of macroeconomic factors such as interest rates. However, it is now considered to be a good measure of intellectual capital. If the value of q is less than 1, i.e. an asset is worth less than the cost of replacing it, then it is unlikely that the company will buy more of that type of asset. If the quotient q (similar to the 'market-to-book value' ratio) is greater than 1, the company has the ability to make unusually high profits because it has something unique that others do not have. That uniqueness is considered to be its people, business systems and processes and customer relationships, which make it different from other companies, i.e. intellectual capital.

9.5.12   Intellectual Capital (IC) index

The Intellectual Capital (IC) index (Intellectual Capital Services, 2003) is an attempt to consolidate all the different indicators into a single index.

Table 9.5 Measurement application tools

Type of tool	Brief description/ focus of tool	Links to strategic objectives	Application area		Measurement approach	Advantages and/or limitations
			Knowledge assets (stocks)	KM initiatives (flows)		
Skandia Navigator	Five areas: financial, customer, process, renewal and development, and human capital, with metrics for each area	Strong	✓		Metrics	Easy to implement, but using too many measures can make it confusing. Does not deal explicitly with the quantification of the value and benefits of KM
Intangible Assets Monitor	Three families of intangible assets: external structure, internal structure and individual competence	Strong	✓		Metrics	Easy to implement, but using too many measures can make it confusing. Does not deal explicitly with the quantification of the value and benefits of KM
Intellectual Capital (IC) index	Consolidates different indicators into a single index and correlates the changes in IC with changes in the market	Strong	✓		Metrics Market value	Easy to implement, but problems with obtaining standardised metrics and industry benchmarking
KM Benefits Tree	Relates knowledge benefits to organisational benefits	Strong		✓	Metrics Economic	Focused on three levels of benefits: knowledge, intermediate and organisational benefits

Approach	Description					Comments
Degussa–Huls approach	Relates KM initiatives to transfer effects and successes based on six dimensions (people, management, processes, technology, innovation, customers and market)	Strong	✓	✓	Metrics Economic	Focused on the impact of key measures and provides an assessment of the impact on business results
IMPaKT Assessor	Relates KM initiatives to performance metrics, strategic objectives and provides for quantification of value	Strong	✓		Metrics Economic	Focused only on measures related to strategic objectives and provides an assessment of the impact on key measures and business. Also facilitates the diagnostic of KM problems and the development of KM initiatives
Market-to-book-value ratio	Based on the concept that intellectual capital can explain the difference between the value assigned by the stock market and its book value	Weak	✓		Market value	Provides a very good overall indicator of value subject to stock market volatility. Limited information for developing strategies on knowledge assets and KM
Tobin's q ratio	Compares the market value of an asset with its replacement cost (book value)	Weak	✓		Market value	Provides a very good indicator of value, but limited information for developing strategies on knowledge assets and KM

The method is based on the principle that movement in the IC index is a reflection of the changes in the market value of an organisation. The IC index was originally developed by Johan and Goran Roos (Roos and Roos, 1997; Roos *et al.*, 1998) and initially applied in Skandia in 1997 to supplement its annual report. The index identifies various categories of intellectual capital: customer, human and structural capital. The company's strategic goals and the nature of the business sector it operates in determine the composition and relative importance or weights assigned to each indicator in the IC index.

9.5.13   *Comparison of selected measurement tools*

Table 9.5 is a comparison of selected application tools with respect to their key attributes such as their link to strategic objectives, principal application and measurement approaches incorporated.

9.6   Conclusions

Measurement is central to improving the value of KM programmes and knowledge assets in organisations. However, it is one of the least developed areas of KM, given the complexities involved in understanding the dynamics of knowledge and its management. Public companies may need to justify investment at any stage because of the need to create and demonstrate its value to shareholders and other stakeholders. Other companies may not be under the same pressure, but the expectation to show success is the same. A performance-based approach is, therefore, vital if KM is to be successfully implemented, evaluated and reviewed. Appropriate methods need to be put in place to monitor and communicate the benefits of KM initiatives, publicising the results to help increase the level of awareness and maintain the level of support and enthusiasm.

Performance measurement of KM and associated knowledge assets is a rapidly evolving area and a variety of measurement approaches and application tools have emerged that could be selected. These range from simple tools incorporating metrics to track the performance of knowledge assets to more advanced tools facilitating the computation of the economic returns on KM and knowledge assets. It is recognised that at the lower levels of maturity, basic metrics to monitor and review the KM strategy may be all that is necessary. However, as an organisation progresses to a stage where implementation of KM is mature and well co-ordinated, a robust measurement system may be required.

This chapter has discussed the different types of performance measures and application tools that could be used in implementing and monitoring a KM strategy. The following chapter discusses issues relating to the development of a KM strategy using the CLEVER approach.

References

Bontis, N. (2000) *Assessing Knowledge Assets: A Review of the Models Used to Measure Intellectual Capital*. Queens Management Research for Knowledge-Based Enterprises, Queens School of Business, Queens University, Ontario, Canada.

Bontis, N., Dragonetti, N.C., Jacobson, K. and Roos, G. (1999) The knowledge toolbox: a review of the tools available to measure and manage intangible resources. *European Management Journal*, **17**(4), 391–404.

Bunz, A.P. and Kirch-Verfuss, G. (2001) Mapping the value of knowledge management. *Knowledge Management Magazine*, **4**(4), 20–25.

Carrillo, P.M., Robinson, H.S., Anumba, C.J. and Al-Ghassani, A.M. (2003) IMPaKT: a framework for linking knowledge management to business performance. *Electronic Journal of Knowledge Management (EJKM)*, **1**(1), 1–12.

Demarest, M. (1997) Understanding knowledge management. *Long Range Planning*, **30**(3), 374–84.

Department of Navy (DON) (2001) *Metrics Guide for Knowledge Management Initiatives*. Available at https:www.doncio.navy.mil (accessed March 2003).

DiPiazza, S.A. (Jr) and Eccles, R.G. (2002) *Building Public Trust: The Future of Corporate Reporting*. John Wiley and Sons, Chichester.

Edvinsson, L. (1997) Developing intellectual capital at Skandia. *Long Range Planning*, **30**(3), 366–73.

EFQM (1999) *Eight Essentials of Excellence: The Fundamental Concepts and Their Benefits*. European Foundation for Quality Management, Brussels, Belgium.

Intellectual Capital Services (2003) http://www.intcap.com (accessed March 2003). Intellectual Capital Services Ltd, London.

Jordan, J. and Jones, P. (1997) Assessing your company's knowledge management style. *Long Range Planning*, **30**(3), 392–8.

Kaplan, R.S. and Norton, D.P. (1996) The balanced scorecard – measures that drive performance. *Harvard Business Review*, **70**(1), 71–9.

Robinson, H.S., Carrillo, P.M., Anumba, C.J. and Al-Ghassani, A.M. (2002) Evaluating knowledge management strategies: an IMPaKT assessment. In: *Proceedings of the 3rd European Conference on Knowledge Management (ECKM 2002)*, Trinity College Dublin, Ireland, 24–25 Sept, pp. 586–98.

Roos, J. & Roos, G. (1997) Valuing intellectual capital. *FT Mastering Management*, **3**, July–Aug, 6–10.

Roos, J., Roos, G., Dragonetti, N.C. and Edvinsson, L. (1998) *Intellectual Capital: Navigating in the New Business Landscape*. New York University Press, New York.

Sheehan, T. (2000) Building on knowledge practices at Arup. *Knowledge Management Review*, **3**(5), 12–15.

Skyrme, D. (1998) *Measuring the Value of Knowledge: Metrics for Knowledge-Based Business*. Business Intelligence, London.

Skyrme Associates (2003) *Knowledge Management Benefits Tree*. Available at http://www.skyrme.com/tools/bentree.htm (accessed March 2003).

Stewart, T.A. (1997) *Intellectual Capital: The New Wealth of Organisations*. Doubleday, New York.

Sveiby, K.E. (1997a) The Intangible Assets Monitor. *Journal of Human Resource Costing and Accounting*, **2**(1), 25–36.

Sveiby, K.E. (1997b) *The New Organizational Wealth: Managing and Measuring Knowledge-Based Assets*. Berrett-Koehler, San Francisco.

Terrett, A. (1998) Knowledge management and the law firm. *Journal of Knowledge Management*, **2**(1), 67–76.

10 Knowledge Management Strategy Development: A CLEVER Approach

Chimay J. Anumba, John M. Kamara and
Patricia M. Carrillo

10.1 Introduction

Previous chapters have considered the nature and dimensions of knowledge management (KM) and have established the case for the effective management of knowledge in the construction industry. It is evident that the successful implementation of KM initiatives requires dedicated effort on the part of managers, which can be in the form of clearly defined strategies.

A strategy is defined as 'a detailed plan for achieving success in situations such as war, politics, business, industry or sport, or the skill of planning for such situations' (CID, 2003). Effective business strategies ensure that plans are in line with organisational goals, gain continuous commitment from top management, allocate enough resources and optimise their use, allow for compatibility between existing and required structures of culture and technology, and increase the likelihood of success (Al-Ghassani *et al.*, 2002).

Strategies for the management of knowledge can be described as 'supply-driven' or 'demand-driven' (Scarbrough *et al.*, 1999). Supply-driven initiatives assume that the fundamental problem of KM is to do with the flow of knowledge and information within the organisation. The aim is to increase that flow by capturing, codifying and transmitting knowledge. There is a tendency for supply-driven initiatives to have a strong technology component. Demand-driven approaches are more concerned with users' perspectives and their motivation and attitudes are seen as important. Strategies within this category usually include reward systems and ways of encouraging knowledge sharing (Scarbrough *et al.*, 1999).

KM strategies can also be described as either 'mechanistic' or 'organic' with respect to the emphasis on either 'explicit knowledge' (mechanistic) or 'tacit knowledge' (organic) (Kamara *et al.*, 2002). Like supply-driven approaches, mechanistic strategies are technology-reliant and focus on the codification of knowledge through knowledge-based expert systems and the use of other information and communication technologies (ICTs). Within the 'organic' domain, techniques such as 'story-telling' and 'communities of practice' are used to capture and transfer tacit knowledge. Snowden (1999) suggests that story-telling can be used to create a self-aware descriptive capability in organisations and to initiate and sustain interventions that create resilience, robustness and redundancy. A 'community of practice,' on the other hand, is an environment where newcomers learn from old-timers by being allowed to participate in certain tasks relating to the practice of the community (e.g. a community of bricklayers) (Hildreth *et al.*, 2000).

Whatever the emphasis, however, two key factors – 'content' and 'context' – determine the success of KM strategies (Kamara *et al.*, 2002). 'Content' refers to the knowledge that is to be managed. 'Context' refers to the organisational setting for the application of knowledge, and includes culture and human dimensions of KM. The relevance and interrelationship between 'content' and 'context' is related to the emphasis (with respect to KM) on knowledge that contributes to business excellence. That is, knowledge that is seen as an essential component (a definable resource) to the performance of specific tasks within organisations. Therefore, knowledge (content) on its own is of little benefit unless it has practical use within an organisation.

The importance of content and context in developing KM strategy means that a 'one-size-fits-all' solution to KM problems is unlikely to be successful (Dixon, 2000). The challenge is therefore for individual firms to determine what strategies and tools are appropriate for managing the knowledge that is beneficial to them. Within the AEC (architecture/engineering/construction) industry, the challenge is both at the firm (i.e. individual firms which make up the industry) and at the project level. However, as was demonstrated in Chapter 7, the two are interrelated. The way knowledge is managed at the firm level influences project performance. On the other hand, lessons learned within projects feed back into operational procedures in individual firms.

This chapter addresses a key challenge that organisations face in developing a KM strategy and focuses on the *how* of KM in relation to the development of KM strategies within construction organisations (the AEC industry). Following the Introduction, the next section (10.2) provides the background to the CLEVER project, an EPSRC (Engineering and Physical Sciences Research Council, UK)-funded research project on cross-sectoral learning in the virtual enterprise (CLEVER). Section 10.3 describes the CLEVER framework for selecting a KM strategy that takes into consideration the business goals, and the specific organisational and cultural context within which the organisations operate. This is followed by Section 10.4 which discusses the practical application of the framework and prototype application, including the findings from workshops on its use in various organisations. Section 10.5 concludes with comments on the wider implications of tools like CLEVER in the management of knowledge in the construction industry.

10.2   The CLEVER project

The intention of this project was to explore the characteristics of KM in different industry sectors in order to derive a cross-sectoral framework that helps companies in different sectors to select KM processes best suited to their circumstances. This was in recognition of the fact that in multi-project environments, the management of project knowledge (i.e. its collection, propagation, reuse and maintenance) is generally accepted as being open to considerable improvement, both within companies and between companies in the supply chain (Siemieniuch and Sinclair, 1993, 1999). The CLEVER project, therefore, focused on three levels: intraproject (especially across departmental, functional and organisational boundaries along the supply chain); interproject (both between other projects and within the wider organisation providing support to projects); and cross-sectoral (for long-term learning opportunities for a range of industry sectors). A novel aspect of the approach adopted for the project was its concentration on organisational and contextual aspects, in place of the usual emphasis on information technology and its implementation. The specific aims of the project were:

- to generate an 'as-is' representation of KM practices in project environments both within and across enterprises in the manufacturing and construction sectors
- to derive generic structures for these practices by cross-sectoral comparisons
- to develop a viable framework for KM in a multi-project environment, within a supply chain context, together with requirements for support
- to evaluate the framework using real-life projects and scenarios supplied by the participating companies.

10.2.1   Project strategy and methodology

The methodology adopted to achieve the project objectives involved continuous review of the academic, industrial and web-based literature; early development of a common research framework within the academic team; and the adoption of a user-centred approach to the classification of current practice and the identification of use cases. This included case studies of industry practice and interviews with different classes of stakeholders both within projects and at senior levels. A prototyping approach was adopted for the development of the generic KM framework, which was then subjected to iterative user evaluation using simulated and real situations.

The basic strategy adopted was to investigate, through case studies, KM practices in the manufacturing and construction sectors with a view to facilitating cross-sectoral learning to the mutual benefit of both sectors. The methodology involved a continuous review of literature, the development of a common theoretical framework for project research within the academic team, and the adoption of a user-centred approach (i.e. continuous input from industrial collaborators) to the classification of current practice and the development of the framework for knowledge transfer.

Theoretical framework for case studies

Figure 10.1 shows the theoretical framework that underpinned the research and the 'as-is' studies of KM. It reflects the understanding of KM and knowledge described above. The interrelationships between the four elements in the diagram illustrate the fact that KM is not usually a linear process of creation, capture, storage, transfer, reuse, etc. that available literature on the subject might suggest. The 'knowledge base' (used in a wider sense) refers to the kind of information, data or project knowledge that are to be managed (its nature, location, etc.). 'Knowledge management processes' refer to the tasks and activities that are implemented to manage knowledge, within the context of a project and/or organisation ('process-shaping factors'). 'Performance measurement' deals with the assessment of the real-time usefulness of KM efforts, since KM is not an end in itself, but a means to achieve business goals.

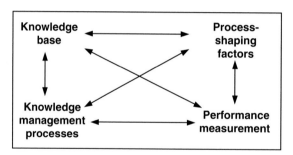

Figure 10.1　Framework for KM research

10.2.2　Studies of KM processes

The KM practices in 15 firms in the construction and manufacturing industries were studied. The studies were based on semi-structured interviews, which lasted for about two hours, with individuals (between one and seven) in each firm. Questions asked revolved around the following themes: (1) the organisational context for the management of project knowledge; (2) the transfer of knowledge between projects (type of knowledge and current processes); and (3) the challenges and opportunities for cross-project KM. Table 10.1 shows the firms that were involved in the research.

The studies on current KM practice revealed that in both the manufacturing and construction sectors, the management of knowledge was characterised by the absence of formal, proactive KM processes. However, there are some examples of good practice in the activities that contributed to

Table 10.1   Firms involved in the CLEVER project

Sector	Type of firm	No. of firms	No. of employees	No. of interviews
AEC	Construction firms	2	Between 1000 and 5000	11
	Construction client	1	About 10 000	5
	Architecture/engineering consultants	2	About 6000	6
Manufacturing	General engineering, aerospace products and services	10	Range from <100 to >1000	10
	Total firms	**15**	**Total interviews**	**32**

the management of project knowledge, for example in the effective use of project management tools, documentation systems and regular revisions of project plans to take into consideration any lessons learnt from past activities. The use of certain procurement options in construction firms also provide an appropriate organisational framework through which people can reuse their tacit learning on subsequent projects. There are also examples of good practice in integrating IT systems and processes to ensure that there is consistency in KM-related activities across an organisation. However, the absence of dedicated KM strategies means that organisations do not derive as much benefit from KM as they should. The results therefore suggest that companies might need assistance in the following areas: (1) identification of their high-grade knowledge; (2) assistance in making high-grade knowledge explicit and highly controlled; and (3) assistance in selecting appropriate strategies for KM that reflect the unique features of their organisations. These findings were taken into consideration in the development of the CLEVER framework.

10.3   The CLEVER framework

The CLEVER Research Team and the Project Steering Committee (PSE) held quarterly meetings to develop the framework over the 20-month research period. Initial technical workshops were held with several of the collaborating companies to present and test the problem-solving aspects of the framework. This iterative process resulted in the CLEVER framework, whose aims are as follows:

- to clarify a 'vague' KM problem(s) into a set of specific KM issues, set within a business context *in order to*
- provide an appropriate and relevant process(es) to solve the identified KM problems *by*

Table 10.2   Stages in the CLEVER framework

Stage	Aim	Outcomes
Define KM problem	To define the overall KM problem within a business context	• Clarification of the KM problem • Distillation of a set of KM issues from the overall problem
Identify 'to-be' solution	To identify required status on a range of knowledge dimensions and to highlight areas of future focus	• Set of concerns or specific KM components of the overall problem on which focus is required
Identify critical migration paths	To identify critical migration paths for each specific KM problem (or dimension of interest)	• Set of key migration paths for each specific KM problem • Overall set of migration paths for the whole KM problem
Select appropriate KM process(es)	To help in selecting the appropriate KM process to move along each migration path	• Set of appropriate KM process(es), which, when tailored to a particular organisation's needs, will address the stated KM problem

- defining the KM problem and linking it to business drivers/goals *and*
- creating the desired characteristics of the 'to be' KM solution *and*
- identifying the critical migration paths to achieve the 'to be' model *and*
- selecting appropriate KM process(es) to use on those paths.

The framework is split into four main stages (Table 10.2) which take the user from an initial definition of a knowledge problem, an identification of where he/she wishes to get to, including the critical migration paths required, through to the provision of appropriate KM processes to aid in the resolution of the user's knowledge problem. The features and application of each stage in the framework are described below.

10.3.1   Define KM problem

The aim of this stage is to define the overall KM problem within a business context, and it involves a description of the perceived problem and identifying the business drivers underpinning it. The characteristics of the knowledge under consideration are defined and the potential users and sources of this knowledge are identified. The probable enablers and inhibitors for identified users and sources and the potentially relevant KM processes (e.g. creation and transfer of knowledge) are also identified. The output of this stage is a clarified KM problem and a set of KM issues emanating from the problem.

The implementation of this stage requires the use of a Problem Definition Template (PDT). The PDT consists of a structured set of questions which are divided into five sections:

(1) type of knowledge
(2) characteristics of knowledge
(3) sources and users of knowledge
(4) current processes
(5) restatement of problem.

These questions assist users to 'think through' a KM problem within their organisation, and are not designed to elicit 'precise' information for quantitative analysis. Tables 10.3 and 10.4 show the first and second sections of the PDT (respectively), which elicit information about the type and

Table 10.3 Section A of the PDT (type of knowledge)

A1: What knowledge are you interested in?						
A2: Please select from the adjacent list, the class(es) of knowledge that best describes this knowledge	(a) Best practice			(b) Equipment/tools		
	(c) Product knowledge			(d) Quality standards/processes		
	(e) Operational process/procedures			(f) Domain/function knowledge		
	(g) Support process/procedures			(h) Human resources		
	(i) Strategies/policies			(j) *Other (please specify)*		
	(k) Control procedures					

	Category of driver	*Business driver*	KM process				
			Knowledge generation	Knowledge propagation	Knowledge transfer	Knowledge location and access	Knowledge maintenance/ modification
A3: What are the business drivers for this knowledge problem?	Structural change	Expansion					
		Restructuring					
		Merger and acquisition					
		Down-sizing					
	(Other)						
	External change	New market					
		New technology					
	(Other)						
	Continuous improvement	Performance improvement					
	(Other)						

Table 10.4 Section B of the PDT (characteristics of knowledge)

B1: What are the characteristics of this knowledge? *(Indicate in the sliding scale how best this knowledge is characterised)*	*Is the knowledge generally:*

Is the knowledge generally:

Explicit: can be captured, codified and formalised — ←explicit [| | | |] tacit→ — **Tacit**: (experience) usually in people's heads

Auxiliary: often general knowledge; never necessary in isolation — ←auxillary [| | | |] critical→ — **Critical**: core to operational effectiveness and achievement of business goals

Discipline based: emphasis on developing single discipline expertise — ←discipline [| | | |] project→ — **Project based**: emphasis on developing multi-disciplinary expertise

Slow change: tends to evolve rather than increase rapidly — ←slow change [| | | |] rapid change→ — **Rapid change**: frequent generation of new or amended knowledge

Other (please specify) ← [| | | |] →

B2: Where is it located?

Is the knowledge mostly:

External: knowledge exists outside the organisation, e.g. it may be bought in — ←external [| | | |] internal→ — **Internal**: knowledge exists within the organisation, tends to be owned

Individual: knowledge held by individual(s) — ←individual [| | | |] shared→ — **Shared**: knowledge is shared and available across the organisation

Specific to problem: knowledge relates to defined problem context — ←specific [| | | |] generic→ — **Generic**: knowledge can be applied across a range of project contexts

Other (please specify) ← [| | | |] →

B3: How is this knowledge acquired?

Is acquisition (learning) mostly by:

Learn by training: knowledge gained by formal training or action on task or tool — ←formal [| | | |] informal→ — **Learn by interaction:** knowledge gained by interpersonal relationships; (in)formally

Other (please specify) ← [| | | |] →

characteristics of the knowledge under consideration (which is a problem) and the business drivers underpinning it.

The first question in the first column of Table 10.3 (A1: What knowledge are you interested in?) refers to the area/activity/process/etc. in which an organisation is experiencing KM problems (e.g. lessons learnt from projects, etc.). The next question (A2) asks the user to select the class(es) of knowledge that best describe the knowledge under consideration. The third question (A3) tries to link the knowledge problem to organisational goals. It seeks to discover the possible origin of the problem. Is it because of structural changes (e.g. expansion) within the company, or external factors, or the need for continuous improvement? If any or all of these factors have resulted in KM problems within the organisation, which KM process(es) are affected? For example, expansion of the company (structural change) may have resulted in problems in the way knowledge is transferred between different departments. If this is the case, a tick is put at the intersection of the 'Expansion' row and the 'Knowledge transfer' column. Depending on the nature of the problem under consideration, more than one 'Category of driver', 'Business driver' or 'KM process' can be selected.

The second section of the PDT, which deals with the characteristics of the knowledge under consideration, is shown in Table 10.4. A set of common scales for describing knowledge is provided, but there is also scope for the user to add his/her own. On each of the five point scales, the user is asked to indicate where the knowledge under consideration is positioned by placing a tick at the appropriate point. For example, for 'lessons learned about projects' knowledge, which is generally held as experience, a tick will be put towards the tacit side of the explicit–tacit scale. If the knowledge is a mixture of both types, then the tick is put in the middle. Question B2 deals with the location of the knowledge, i.e. where it is held/stored, or where it can be obtained/accessed from. Question B3 deals with the acquisition of knowledge by people *within* the organisation. That is, how does one acquire the knowledge under consideration? Acquisition of knowledge from outside the organisation (e.g. through recruitment) is not considered since the focus is on the KM problems with respect to the business context of the organisation (or sub-unit as appropriate).

Other sections (C, D and E) of the PDT ('sources and users of knowledge', 'current processes' and 'restatement of problem') provide further analysis of the problem to enable the user (or users) to develop a clear and agreed statement of the KM problem.

10.3.2   *Identify 'to-be' solution*

This stage highlights the problem areas the user wishes to focus on. It is used to confirm the characteristics of the current 'as-is' position and identify the desired ('to-be') position on each problem area with regard to organisational strategy and policy. A clear set of concerns are extracted and prioritised, and

Table 10.5 Knowledge dimensions guide

Left anchor	Continuum	Right anchor
Explicit: automated/process-based decision making	F C FC *(Approach to decision making)*	**Tacit**: human-based decision making by discussion/consensus
Auxiliary: focus on performance, efficiency and costs	FC *(Recognising core competence)*	**Critical**: focus on knowledge as a competitive edge
Discipline based: emphasis on developing single-discipline knowledge domains	FC *(Openness to change/flexibility)*	**Project-based**: focus on developing multi-disciplinary project knowledge
Slow change: competitive edge depends on efficiency of knowledge	FC FC *(Requirement to innovate)*	**Rapid change**: competitive edge depends on ability to innovate
External: emphasis is on managing knowledge which can be bought in as required	FC *(Knowledge ownership and availability)*	**Internal**: emphasis is on owning knowledge that is particularly rare or valuable
Individual: having access to the knowledge is more important than sharing it	C F *(Knowledge as an organisational asset)*	**Shared**: knowledge is seen as an organisational asset to provide added value
Problem-specific: excellent for recurrent problems (runners and repeaters)	C F *(Reuse of knowledge)*	**Generic**: reuse of knowledge is important; allows wider allocation of work
Learn by training: ensure technical/professional competence	CF *(Propagation of organisation's culture)*	**Learn by interaction**: assists transfer of organisational values

these are used to identify migration paths for each identified problem. The output of this stage is a set of concerns or specific KM components of the overall problem that the user wishes to focus on.

This stage is supported by a 'knowledge dimensions guide' (Table 10.5), which is a sliding scale with predefined 'states' on which the user can indicate his/her current ('as-is') and desired future ('to-be') KM positions with a 'C' and 'F' respectively. Detailed explanation of each of the dimensions is provided to assist the user to more accurately define his/her current and desired positions. For example, implications of 'individual/shared' continuum (Table 10.5) are provided in Table 10.6. From these identified positions, the critical areas of concern are identified. These are the areas where the desired position is furthest from the current situation. From the example given in Table 10.5, these critical areas include the transfer of tacit knowledge to explicit knowledge to aid decision-making efficiency and the sharing of knowledge to ensure wider use as an organisational asset. However since the gap between the current and desired states is widest in the 'individual/shared' continuum, this problem can be considered to be of higher priority than the need to transfer knowledge from tacit to explicit.

10.3.3 *Identify critical migration paths*

This stage focuses on defining how the user wishes to proceed from the current ('as-is') situation to the desired ('to-be') position. A set of predefined 'squares' that relate to each problem identified in the previous stage are selected. The user then maps out his/her current situation, where he/she wants to be, and the path he/she wants to follow. Each identified problem is considered in turn, and the overall set of migration paths are mapped out for the overall KM problem under consideration.

Table 10.6   Implications for adding value of the class of (individual/shared) knowledge

Individual (access and use important)	Shared (sharing and use important)
• Knowledge is an underused asset for the organisation	• Potential for reuse and organisational learning adds value
• Organisation does not own this knowledge – can walk out of the door	• Organisation tends to control this asset
• Can be a transient asset with limited reuse	• Filtering and accessibility are important
• Tends to add value via local application	• Often enables added value by application of lessons learnt
• Knowledge often not available for general reuse	• Reallocation of resources is easier
• Can often be task-focused with narrow perspective	• Can be an excellent change enabler – common models
• Can be a scare resource and costly to replace	• Potential source of/barrier to organisational culture
• Conflicts can lead to paralysis and lost opportunities	• Shared knowledge provides greater flexibility in thinking

For this stage in the framework, a set of 'migration path tools' is used. These are a set of matrices ('squares') that define the possible implications of migrating from the current to the desired knowledge solution. For example, if the sharing of knowledge is considered, this applies to 'knowledge as an organisational asset' (Table 10.5). One of the 'squares' that would be applicable to this situation is shown in Figure 10.2. The desired position on the 'square' is the top left-hand corner (shared internal/explicit), and this can be achieved by 'migrating' from the bottom right of the 'square' in the directions indicated. The decision on which path to follow will depend on the resources of the organisation.

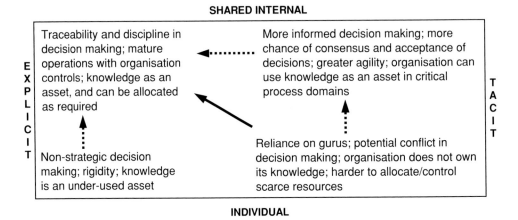

Figure 10.2   Migration path tool for 'individual/shared' and 'explicit/tacit' knowledge

10.3.4   Select appropriate KM process(es)

This stage deals with the selection of appropriate KM process(es) to move along each migration path. Thus for each migration path defined in the previous stage, the relevant KM process is selected from a standard list of processes. Organisational enablers/resistors that may facilitate or inhibit the implementation of the selected process are also identified. This will enable the organisation to develop specific plans (based on enablers/resistors) to implement the selected strategies that relate to its stated KM problem.

The selection stage uses 'generic KM process models'. For each migration path, a generic process (e.g. propagate/transfer knowledge) is selected. The possible factors that could facilitate ('enablers') or hinder ('resistors') the migration to the desired 'to-be' situation are also identified. Figure 10.3 shows the generic process model for the transfer of knowledge, which might be appropriate for migrating from knowledge from 'tacit/individual' to 'explicit/shared internal' (see Figure 10.2). There are four steps in this process: 'identify knowledge to be transferred', 'identify knowledge sources',

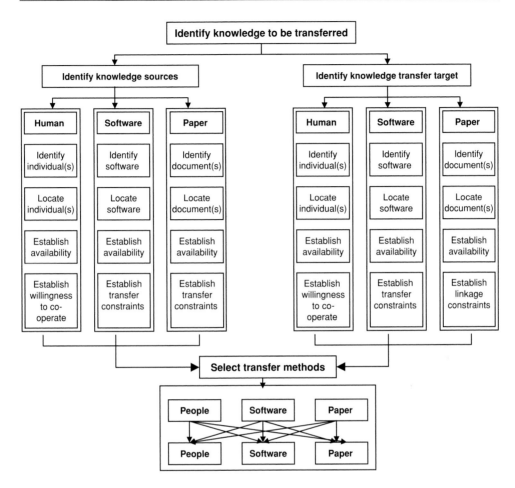

Figure 10.3   Generic process model for knowledge transfer

'identify knowledge transfer target' and 'select transfer methods'. A clear identification of the 'source' and 'destination' of knowledge to be transferred will determine whether it is people-to-people transfer, people-to-paper transfer, etc. Resistors and enablers that affect each step should be identified. For example, possible resistors to the 'identify knowledge sources' could be: difficulty in identifying the sources of knowledge, or the fact that the knowledge to be transferred is not 'visible' to decision makers. On the other hand, enablers to this step could be clear identification of knowledge sources, and a willingness to share by human repositories of knowledge. Transfer methods (e.g. Figure 10.4) are based on established methods such as mentoring, story-telling, etc. However, by distilling a KM problem, identifying migration paths and the organisational resistors and enablers that might affect a selected KM process, the user is assisted in selecting an appropriate strategy for its solution that reflects his/her organisational context.

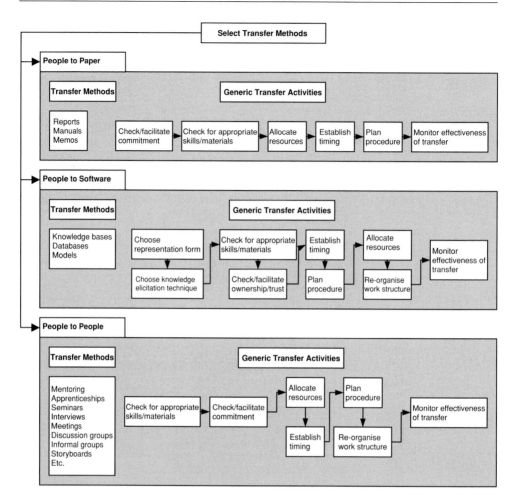

Figure 10.4 Detailed knowledge transfer processes (people–paper; people–software; people–people)

10.4 Utilisation and evaluation of the framework

The CLEVER framework can be used to analyse the KM problem for an organisation or a unit within that organisation (e.g. subsidiary, department, business unit, etc.). It can be completed by one individual. However, if the problem under consideration is organisation-wide, it is helpful if several individuals (especially managers and users of knowledge) are involved in its completion. A group of people can use the framework individually and then discuss the responses of each person. In this case the type of knowledge being considered (question A1 of the PDT – Table 10.3) should be specified so that everybody has the same point of reference. Alternatively, the framework can be used collectively by a group, with somebody acting as facilitator. When using the framework, it is helpful to consider *one* KM problem at a time. This will provide a central focus for all the stages in the framework.

The use of the CLEVER framework has been evaluated by four organisations (three construction firms and one IT company) through specific (half- to two-day) workshops organised (between April and December 2001) and facilitated by the research team. Participants from each of these firms were asked to identify a real KM problem within their organisation, and then the CLEVER framework was used to analyse the problem and select KM strategies appropriate to their contexts. Feedback was provided during and after the workshop through informal comments and an evaluation questionnaire respectively. The questionnaire contained questions relating to the ease of use, layout, relevance to the business, and the appropriateness of the verbal and written guidance on the use of the framework. Participants were also asked to comment on the usefulness of the framework, the flow between the four stages and on the framework generally.

The results of the evaluation workshops show that the CLEVER framework is very useful for encouraging companies to consider, in a consensus manner, the extent of their KM problems. Working through the four stages of the framework produces a more enlightened view of KM, a prioritisation of problems with associated resistors and enablers and a well-defined process to eliminate the KM problem. This is regarded as of immense value to any organisation. However, in its paper version, the framework has problems that need to be addressed. These are the format, the need for a facilitator and the wording used in the migration paths tool. Essentially, the content is right, but the format needs amending.

Another issue that arose from the framework and the evaluations was the need to assess the readiness of an organisation for KM. Some provision for this has been made in the CLEVER framework, but there is scope for a dedicated project to develop an appropriate KM readiness assessment model.

To address some of the limitations of the paper-based version of the CLEVER framework, a prototype software (CleverKM™) has been developed. Figures 10.5 to 10.9 show screen shots of the system.

10.5 Conclusions

The CLEVER framework described in this chapter focuses on the definition and analysis of a knowledge problem in order to facilitate the selection of an appropriate strategy for the KM within an organisation. The rationale underpinning the framework is based on a number of issues arising from the literature and the 'as-is' studies of KM in the collaborating organisations.

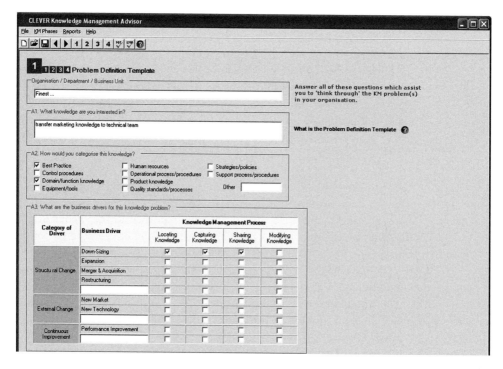

Figure 10.5   CLEVER problem definition template

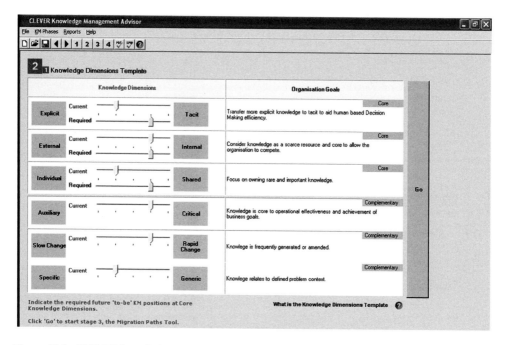

Figure 10.6   CLEVER knowledge dimensions template

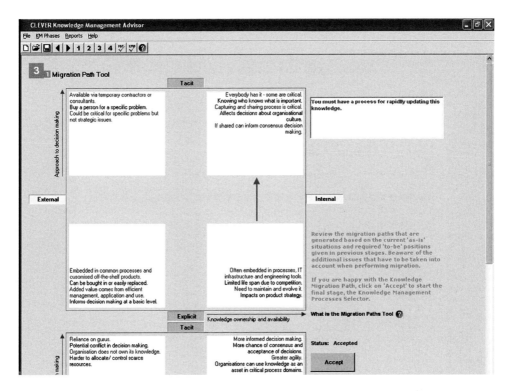

Figure 10.7 CLEVER migration path tools

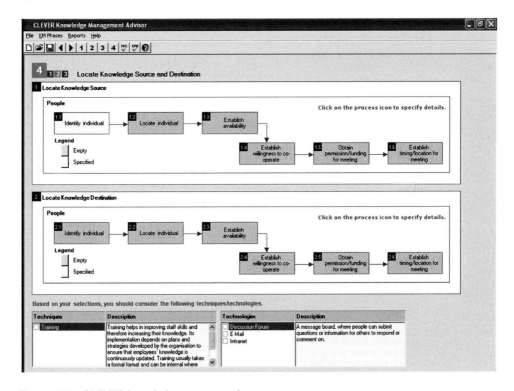

Figure 10.8 CLEVER knowledge process selector

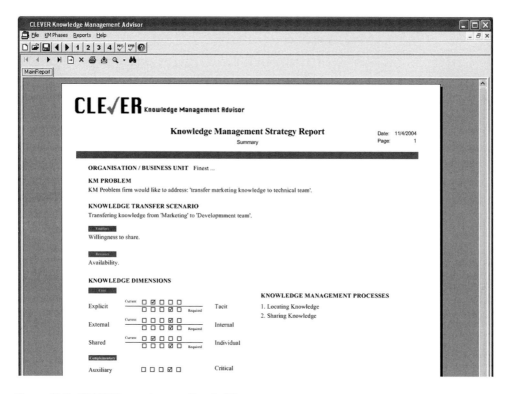

Figure 10.9 CLEVER report generation facility

First, the management of knowledge is not an end in itself, but a process, which is aimed at creating value, increasing productivity and gaining/sustaining competitive advantage (Siemieniuch and Sinclair, 1999). The framework was therefore linked to the business drivers for change within an organisation. The emphasis on business drivers also underscores the operational understanding of knowledge used in the research; that is, the relevance of knowledge within its context of use.

Second, it was acknowledged that there are already many KM 'solutions' relating to the various processes of knowledge generation, capture, transfer, etc. The focus was therefore not on developing another 'process', since most problems in implementing KM strategies within organisations are not necessarily because of these solutions *per se*, but owing to their inappropriate use within the *context* of these organisations (Storey and Barnet, 2000). What is important, therefore, is an adequate understanding of the problem, within the context of an organisation, and the potential enablers and resistors that will affect its resolution. However, because KM problems are multi-dimensional, an expert system that is based on 'if-then-else' constructs would be inappropriate to address them.

Further work on automating the framework, which is under way, will enhance its usefulness in clearly linking KM issues to the business goals of an organisation, thereby ensuring that the effective management of knowledge will lead to increased value, productivity and competitive advantage for the company implementing it.

References

Al-Ghassani, A.M., Kamara, J.M., Anumba, C.J. and Carrillo, P.M. (2002) A tool for developing knowledge management strategies. *Electronic Journal of Information Technology in Construction (Special Issue on ICT for Knowledge Management in Construction)*, **7**, 68–82 (available at http://www.itcon.org/2002/5/).

CID (2003) Cambridge International Dictionary of English. Available at http://dictionary.cambridge.org/ (accessed February 2003).

Dixon, N.M. (2000) *Common Knowledge: How Companies Thrive by Sharing What They Know*. Harvard Business School Press, Boston, Massachusetts.

Hildreth, P., Kimble, C. and Wright, P. (2000) Communities of practice in the distributed international environment. *Journal of Knowledge Management*, **4**(1), 27–38.

Kamara, J.M., Augenbroe, G., Anumba, C.J. and Carrillo, P.M. (2002) Knowledge management in the architecture, engineering and construction industry. *Construction Innovation*, **2**(1), 53–67.

Scarbrough, H., Swan, J. and Preston, J. (1999) *Knowledge Management: A Literature Review*. Institute of Personnel and Development, London.

Siemieniuch, C.E. and Sinclair, M.A. (1993) Implications of CE for organisational knowledge and structure – a European, ergonomics perspective. *Journal of Design and Manufacturing*, **3**(3), 189–200.

Siemieniuch, C.E. and Sinclair, M.A. (1999) Organisational aspects of knowledge lifecycle management in manufacturing. *International Journal of Human–Computer Studies*, **51**, 517–47.

Snowden, D. (1999) Liberating knowledge. In: *Liberating Knowledge* (J. Reeves, ed.), pp. 6–19. Caspian, London.

Storey, J. and Barnet, E. (2000) Knowledge management initiatives: learning from failure. *Journal of Knowledge Management*, **4**(2), 145–56.

11 Corporate Memory

Renate Fruchter and Peter Demian

11.1 Introduction

'If HP would know what HP knows, we would be three times more profitable' said the former chief executive of HP, Lew Platt (Davenport and Prusak, 1998). People and knowledge are a corporation's strategic resources. They constitute in a sense the corporate memory. Knowledge is the only unlimited resource, the one key asset that can grow the more it is used. As Peter Drucker emphasised, 'knowledge is the new basis of competition in post-capitalist society'. The typical *knowledge worker* draws from a vast well of previous projects, design experience and best practices. This can be experience acquired by the individual or by his/her mentors or professional community. Reusing knowledge from past experiences is a natural process, which, if supported, can lead to improved designs and more effective management of constructed facilities. This activity is referred to as *knowledge reuse*. This chapter will focus on design knowledge capture, sharing and reuse from a corporate memory, but the principles discussed apply to any typical activity exercised by knowledge workers. Knowledge reuse often fails for numerous reasons, including:

- Designers do not appreciate the importance of knowledge capture because of the additional overhead required to document their process and rationale. Consequently, knowledge is often not captured.
- Even when knowledge capture does take place, it is limited to formal knowledge (e.g. documents). Contextual or informal knowledge, such as the rationale behind design decisions or the interaction between team members in a design team, is often lost, rendering the captured knowledge not reusable, as is often the case in current industry documentation practices.
- There are no mechanisms from both the information technology and organisational viewpoints for capturing, finding and retrieving reusable knowledge.

Empirical observations of designers at work show that internal knowledge (from the designer's own memory) reuse is effective since: (1) a designer can quickly *find* (mentally) reusable items, and (2) a designer can remember the context of each item, and can therefore *understand* it and reuse it more effectively.

This chapter emphasises the notion of *knowledge in context*. Knowledge in context is design knowledge as it occurs in a designer's personal memory: rich, detailed and contextual. This context includes design evolution (from sketches and back-of-the-envelope calculations to detailed 3D CAD, analysis and simulations), design rationale and relationships between different perspectives within cross-disciplinary design teams.

We define the *corporate memory* as a repository of knowledge in context; in other words, it is an external knowledge repository containing the corporation's past projects that attempts to emulate the characteristics of an internal memory, i.e. rich, detailed and contextual. The corporate memory grows as the design firm works on more projects. This chapter introduces innovative knowledge capture and interaction metaphors and their software implementation that act as a repository of knowledge in context.

Metaphor is used in a human–computer interaction sense. Metaphors increase the usability of user interfaces by supporting understanding by analogy. Metaphors function as natural models, allowing us to take our knowledge of familiar, concrete objects and experiences and use that knowledge to give structure to more abstract concepts. Modern operating systems use the *desktop metaphor*. Online services use shopping cart and checkout metaphors to relate the novel experience of buying online to the familiar experience of buying at a bricks and mortar store (Nelson, 1990).

The conceptual development phase is where some of the most influential design decisions are made in product development. The National Materials Advisory Board (1991) and the Institute for Defense Analysis (Winner *et al.*, 1988) claim that up to 80% of the final project costs are committed during conceptual design. Nevertheless, concepts developed in the early phase of design emerge in informal and unstructured activities, forms and media, which render the archival process difficult.

Several advantages of archiving design information follow. For example, capturing design information allows knowledge to be shared and revisited by the core team, project managers and the client. In addition, a new team member may take a long time to get 'up to speed'. This time may be reduced if design decisions are available, permitting the new worker to quickly understand the context and evolution of the product. More importantly, the notion that all design is redesign drives the need for archiving of design information for the redesign process and should not be limited to personal experience, but take advantage of accumulated knowledge of past designers.

The motivation behind the development of knowledge reuse systems is that the capture and reuse of knowledge is less costly than its recreation. Nevertheless, the state-of-practice to find, retrieve, explore, understand and assess corporate knowledge for use and innovation is difficult, time-consuming and out of context. In many architecture, engineering and construction

(AEC) firms today, knowledge capture and reuse is limited to dealing with paper archives. Even when the archives are digital, they are usually in the form of electronic files (documents) arranged in folders which are difficult to explore and navigate. A typical query might be, 'how did we design previous cooling tower support structures in hotel building projects?' In many cases, the user of such systems is overloaded with information, but with very little context to help him/her decide if and what to reuse. Figure 11.1 illustrates an example of a sub-folder of a digital project folder archive representing the structural drawing documents developed during the schematic design. Interesting to note is the large number of documents in this folder (1201 documents) all having cryptic file names and yielding very little information as to where relevant items to the query could reside.

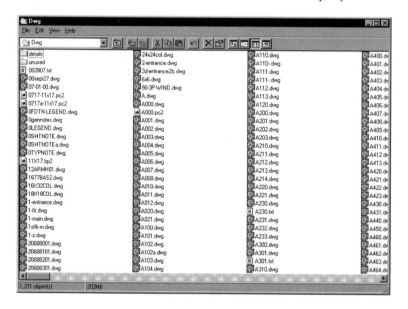

Figure 11.1 Example of a document folder illustrating lack of contextual information in support of design-specific information queries

This chapter addresses the following questions:

- What is the nature of knowledge capture, sharing and reuse?
- What are the key characteristics of the knowledge reuse process?
- How can the design knowledge reuse process in the AEC industry be supported by a computer system?
- What are natural idioms and interaction metaphors that can be modelled into a computer system to provide an effective knowledge reuse experience to a designer?

The objective is to assist the designer and to support the process of design knowledge reuse rather than to automate it.

Following the Introduction, Section 11.2 presents the research methodology, while Section 11.3 discusses findings from related research on design knowledge reuse. Sections 11.4 and 11.5 distinguish between the concept of tacit and explicit knowledge, and discuss their implications in the context of knowledge capture, sharing and reuse. The chapter concludes in Section 11.6 that knowledge reuse often fails, as existing archiving systems do not support designers in reusing knowledge. It is argued that the alternative prototype systems developed (ProMem^{TM} and CoMem^{TM}) could enhance the reuse of knowledge by facilitating the retrieval of relevant items easily, understanding the context and the way the knowledge associated with the items evolved to be able to reuse them effectively.

11.2   Research methodology

The study focuses on how one can capture with high fidelity, and least overhead to the designer, the knowledge experience that constitutes conceptual design. This question addresses two important topics in design capture: what to capture and how to capture. In addressing the first topic, we needed to discover which media types conceptual design activities occur through. By observing how design knowledge is generated, we gain a better understanding to find non-intrusive ways to capture and ultimately reuse this information.

A scenario-based approach was used to analyse the current practice, identify needs, activities, information exchange, interaction scenarios, explore new interaction metaphors and develop the prototype systems that support knowledge, capture, sharing and reuse. The scenario-based design of human–computer interaction (Rosson and Carroll, 2001) was adopted as a working scenario-based methodology. The premise behind scenario-based methods is that descriptions of people using technology are essential in analysing how technology is reshaping their activities. The state-of-practice was observed using ethnographic methods. For a discussion of the use of ethnographic methods for design see Blomberg *et al.* (1993).

The scenario-based design process begins with an analysis of current practice using *problem scenarios*. These are transformed into *activity scenarios*, which are narratives of typical services that users will seek from the system. *Information scenarios* are elaborations of the activity scenarios that provide details of the information that the system will provide to the user. *Interaction scenarios* describe the details of user interaction and feedback. The final stage is prototyping based on the interaction scenarios and evaluation. The process as a whole from problem scenarios to prototype development is iterative.

In the context of this chapter we distinguish between data, information and knowledge in the following ways: data represents raw material, i.e. objective facts captured in structured records; information represents data that have a meaning, relevance, purpose and value for the creator and consumer of specific data. The transition from data to information takes place through contextualisation, categorisation and synthesis of data. Knowledge is created during communicative events within and between people through activities such as conversations, what-if scenarios comparing options, connecting ideas and solutions, analysing consequences of different decisions, exploration, brainstorming, etc. We make an additional distinction between informal and formal communicative events and media. We offer the following working definitions used in this study.

- *Modes of communication* – channels that designers use to communicate, e.g. verbal communication, written communication, gestured communication.
- *Multi-modal* – the combination of two or more modes of communication.
- *Media* – something on which information may be stored, e.g. paper, video, audio.

Communication media may be:

- *Formal* – enables the production of machine-interpretable information that can be quickly and accurately processed.
- *Informal* – does not facilitate capturing and processing of the generated information. In this study we consider captured video/audio as informal media. Although signal processing and voice recognition offer techniques to formalise these media, video/audio captured in its raw state is certainly less formal than digital text.

Design activities may also be:

- *Formal* – these are structured activities for which there are well-defined interaction protocols and archival procedures, such as design development review meetings and design documentation.
- *Informal* – these are unstructured activities that are intended to foster creativity and exploration. These activities do not have well-specified protocols and archival procedures. They tend to occur in an ad hoc fashion during individual or team interaction exploration of the design space.

Design knowledge can be:

- *Tacit* – information that an individual accumulates through experiences. This knowledge resides in the mental model of the individual. Tacit knowledge embodies individual and shared experiences in a variety of circumstances to better capture the context and rationale behind design decisions (Nonaka and Takeuchi, 1995).

- *Explicit* – documented knowledge that can be shared, accessed and re-trieved. The research effort addressed the unexplored aspect of creating an information retrieval system to capture, index and distribute design knowledge. Providing designers with improved access to this knowl-edge and information leverages the experiences of ongoing and previ-ous design efforts with the expectation to improve quality and efficiency of design.

We assert that a primary source of information behind design decisions is embedded within the verbal conversation among designers. Capturing these conversations is difficult because the information exchange is un-structured and spontaneous. In addition, discourse is often multi-modal. It is common to augment speech with sketching and gesturing. Audio/video media can record these activities, but do not provide an efficient means to index the captured information. How can such rich contextual content, i.e. knowledge that constitutes conceptual design generated during informal events such as brainstorming or project review sessions, be captured with high fidelity and least overhead to the team members?

We view knowledge reuse as a step in the knowledge lifecycle (Fruchter and Demian, 2002). Knowledge is created as designers collaborate on design projects using data, information and knowledge. It is captured, indexed and stored in an archive, as will be presented in the next sections. At a later time, it is retrieved from the archive and reused.

Finally, as knowledge is reused it is refined and becomes more valuable. In this sense, the archive system acts as a knowledge refinery (Figure 11.2).

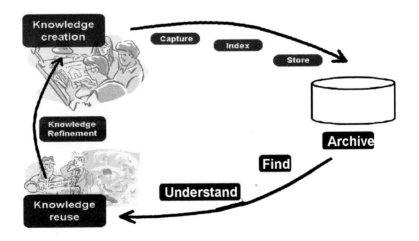

Figure 11.2   Knowledge lifecycle

11.3 Related research

Related research studies on design knowledge reuse focus either on the *cognitive* aspects or on the *computational* aspects. Research into the cognitive aspects of reuse have helped to identify the information needed by designers. Kuffner and Ullman found that the majority of information requested by mechanical engineers was concerning the operation or purpose of a designed object, information that is not typically captured in standard design documents (drawings and specifications) (Kuffner and Ullman, 1990). Finger observed that designers rarely use CAD tools to help them organise and retrieve design information (Finger, 1998). This research extends these findings by formalising the requirements for contextual information when reusing items from previous projects.

The issue of how to capture informal knowledge in project design teams has received extensive attention from researchers in design theory and methodology. The value of contextual design knowledge (process, evolution, rationale) has been repeatedly recognised, but so has the additional overhead required of the designer in order to capture it. The presented research efforts and prototypes build on Donald Schon's concept of the reflective practitioner (Schön, 1983). Researchers in the study of design have focused on either the sketch activity, i.e. learning from sketched accounts of design (Kosslyn, 1981; Olszweski, 1981; Stiedel and Henderson, 1983; Goel, 1995; Tversky, 1999) or verbal accounts of design (Cross and Roozenberg, 1992; Cross, *et al.*, 1996). Some researchers have studied the relation between sketching and talking (Eastman, 1996; Goldschmidt, 1991).

On the computational side, research into design knowledge reuse focuses on design knowledge *representation* and *reasoning*. Knowledge representation ranges from informal classification systems for standard components to more structured design rationale approaches (Hu *et al.*, 2000, give an overview). There is a trade-off in design rationale systems between the overhead for recording design activities and the structure of the knowledge captured.

Highly structured representations of design knowledge can be used for reasoning. However, these approaches usually require manual pre- or post-processing, structuring and indexing of design knowledge. For example, ARCHIE is a case-based reasoning tool for aiding architects during conceptual design (Domeshek and Kolodner, 1993). CASECAD enables designers to retrieve previous design cases based on formal specifications of new design problems (Maher, 1997). IDEAL is a model-based reasoning tool that uses both general domain knowledge as well as knowledge from specific cases (Bhatta *et al.*, 1994). These tools enable knowledge retrieval and reuse based on *a priori* set representations that are specific to narrowly defined domains and media types.

This research brings together the cognitive and computational approaches. The study considers reuse to be a combined effort involving both the human and the computer. Therefore, the issues of design knowledge reuse are addressed as a human–computer interaction (HCI) problem, and the approach takes a user-centred view to designing this interaction. In this approach, capture and indexing take place in real time, with the least possible intrusion on the design process. Knowledge is captured to support the typical communication and co-ordination activities that occur during collaborative design.

11.4   Tacit knowledge capture, sharing and reuse

Design decisions and rationale, in a design team, are most frequently communicated during informal design activities, especially in the conceptual phase. These design activities most often occur through informal media, such as sketching, verbal explanations or gesture language. Unless captured, this knowledge is perishable since only a small fraction of this information makes it into project documentation. By capturing informal design activities in informal media types, design rationale and design decisions are made explicit in project archives that can be shared in real-time or revisited in the future. This type of information will be able to provide context to design decisions and enable future projects to apply the tacit or learned knowledge of the archived group to the current project.

While traditional product documentation captures explicit knowledge such as requirements, specifications and design decisions, often the contextual or tacit knowledge of the design group is lost. Concept generation and development occur most frequently in informal media where design capture tools are the weakest. This statement has strong implications for the capture and reuse of design knowledge because conceptual design generates the majority of initial ideas and directions that guide the course of the project. Sketching is a natural mode for designers, instructors or students to communicate in highly informal activities such as brainstorming sessions, project reviews, lectures or Q&A sessions. Often, the sketch itself is merely the vehicle that spawns discussion about a particular design issue, or it is used to express a new concept. Thus, from a design capture perspective, capture of both the sketch itself and the discussion that provides the context behind the sketch are important. It is interesting to note that today's state-of-practice design experience is not captured and knowledge is lost when the whiteboard is erased or the paper napkin sketch is tossed away.

While common usage of the term discourse implies a 'verbal interchange of ideas', our working definition expands the notion beyond speech and writing to include other modes of communication such as sketching and gesturing. This impacts the process of capturing the discourse.

11.4.1 Scenario

An ethnographic study observed a design team composed of three graduate-level mechanical engineers along with a project sponsor/client and an industry mentor. The goal of the project was to design, prototype and test a pedestrian-safe bumper for automobiles. The design team had nine months and a budget of US $15 000 to complete the project. During the nine months, design information and knowledge was collected in several design activities. These included video and audio records of group meetings, personal notes that were shared on the web, audio records of phone conversations, email interaction, video tape of focus group and interview sessions, memos, presentation materials, and product documentation and reports. This information totalled almost 1000 pages of text, 40 hours of video (with audio) and 15 hours of audio (alone).

11.4.2 Observations

Designers transition through different media. Design decisions and rationale, in a design team, are most frequently communicated during informal design activities, especially in the conceptual phase. These design activities most often occur through informal media (Figure 11.3; noun phrases are directly related to new concepts generated through discourse, decisions or rationale created during the discourse (Mabogunje and Leifer, 1997)). Unless captured, this knowledge is perishable, for we have observed that only a small fraction of this information makes it into project documentation. By

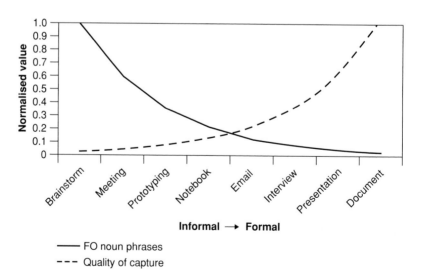

Figure 11.3 Paradox of capturing informal design activities (the information to be captured is highly informal where the design capture tools are weakest)

capturing informal design activities in informal media types, design rationale and design decisions become explicitly stated in project archives that
can be shared in real-time or revisited in the future. This type of information
will be able to give context to design decisions and enable future projects to
apply the tacit or learned knowledge of the archived group to the current
project.

Note that of all the knowledge captured through this study, 52% of all
noun phrases related to generated concepts were found in the least formal
types of media: video and audio. This information is usually lost in the state-
of-practice projects due to the lack of structures and archival procedures for
this type of design activity and information. Formal reports, which are often
the extent of product documentation, only accounted for 5% of the total noun
phrases that were recorded from the design team.

11.4.3 *Needs*

Knowledge management (KM) tools must provide ways of capturing tacit
knowledge because it is rich in design rationale. However, techniques also
need to be developed in structuring this information. Tacit knowledge is
communicated through informal design activities and can be captured in
informal media without imposing any additional burden on the designer.
Since informal media, by definition, is difficult to index, capturing tacit
knowledge in formal media is just the first step. The ability to index this
information and provide quick, relevant access to the information is just as
important.

In conceptual design activity, we have observed that sketching is a natural
mode for designers to communicate in highly informal design activities
such as brainstorming sessions. Often, the sketch itself is merely the vehicle
that spawns discussion about a particular design issue. Thus, from a design
capture perspective, we pose the question: is it more important to capture the
sketch itself or the discussion that provides the context behind the sketch?
Clearly, the discussion surrounding the sketch is valuable to capture. From
the previously mentioned study, we observed that the majority of concepts
are generated and developed through the discussions of highly informal
design activities. On the other hand, the sketch itself is also valuable because
it may act as a mnemonic for discussion around the sketch. It is interesting
to note that in today's common practice, neither of these is captured and
knowledge is lost when the whiteboard is erased.

How does one capture and index informal design activities such as the
design conversations around sketching activity? While this information can
be captured through informal media such as video, it becomes a burden to
index for reuse. For example, indexing videotape would require a time-
stamped transcription. For a tape of an hour's length, this process could
easily take over four hours.

11.4.4   Interaction metaphor and prototype

The metaphors of 'paper and pencil' and 'chalk-talk' were used to develop a prototype tool called RECALL™ (Fruchter and Yen, 2000). RECALL™ is a capture application written in Java that indexes in real-time the discourse and sketch activities. RECALL™ takes advantage of the fact that designers transition through various media and acknowledges the relationship that exists between sketching and the conversations around the drawing including verbal communication and gestures. The drawing application synchronises with audio/video capture and encoding through a client–server architecture. Once the session is complete, the drawing and video information is automatically indexed and published on a web server that allows for distributed and synchronised playback of the drawing session and audio/video from anywhere at anytime. In addition, the user is able to navigate through the session by selecting individual drawing elements as an index into the audio/video and jump to the part of interest (Figure 11.4).

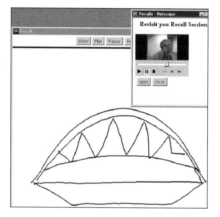

Figure 11.4   Screenshot of RECALL™ playback session. The sketch along with the indexed video and audio can also be replayed

The RECALL™ technology invention is aimed at improving the performance and cost of knowledge capture, sharing and reuse. It provides the following benefits:

- transparent graphical, audio/video indexing
- zero cost and overhead for production, i.e. editing and indexing rich media content
- zero cost and overhead for publishing rich content on the web
- interactive access and retrieval of knowledge and information, i.e. sketch, audio/video on demand.

The RECALL™ has been tested and deployed in different user scenarios in education and industry pilot settings, such as:

- *Individual brainstorming*, where a project team member has a 'conversation with the evolving artifact' enacting Schön's 'reflective practitioner' and using a TabletPC augmented with RECALL™ and then sharing his/her thoughts with the rest of the team by publishing the session on the RECALL™ server.
- *Team brainstorming* and project review sessions, using a SmartBoard augmented with RECALL™.
- *Best practice capture*, where senior experts, such as designer, engineers or builders, capture their expertise for the benefit of the corporation.

The archived RECALL™ sessions become a key part of a corporate memory that facilitates knowledge reuse in context.

The ethnographic studies and the use of the RECALL™ prototype support the hypothesis that concept generation and development occurs most frequently in informal design activities such as brainstorming sessions. While it is possible to record notes and sketches, often the contextual information that is debated verbally is lost. RECALL™ is deployed in an interdisciplinary, geographically distributed project-based engineering class at Stanford University to determine its impact in team interaction, knowledge capture and design reuse (Fruchter, 1999, 2003a, b). By capturing and indexing the informal tacit knowledge of design collaboration, important design decisions that capture not only the final product but also the context and design rationale behind the product can be documented, shared and archived. This allows future users of this knowledge to explore and understand both product and process, i.e. the specific design solution and the rationale and steps towards a decision, respectively.

11.5   Tacit and explicit knowledge capture, sharing and reuse

11.5.1   *Scenario*

An ethnographic study was conducted in a structural design office in California. The objective of the study was to investigate the reuse process qualitatively, and gain a deeper understanding of the steps involved and the types of information reused, rather than to analyse the designers' work practices quantitatively. The firm has three offices in the US with a total of 20 engineers. The California office employs five engineers, including the founder and senior engineer. The study focused on the interactions between an expert (the senior engineer) and a novice structural designer. Internal knowledge reuse was observed and recorded when the novice came to the expert with questions. The observations indicated that the expert always referred to his work on previous projects when answering these questions.

The following is an example of a problem scenario that was developed based on empirical observations.

An expert structural designer, Eric, and a novice, Nick, both work for a structural design office in northern California. The office is part of the 'X Inc' Structural Engineering Firm. They are working on a ten-storey hotel that has a large cooling tower unit. Nick must design the frame that will support this cooling tower. Nick gets stuck and asks Eric for advice. Eric recalls several other hotel projects that were designed by X Inc. He tells Nick to look at the drawings from the Bay Saint Louis project, a hotel project that X Inc designed a couple of years ago. Nick spends over an hour looking for the Bay Saint Louis drawings in the X Inc paper archive. He eventually finds the drawing sheet with the Bay Saint Louis cooling tower frame. He shows it to Eric. The drawing shows the cooling tower frame as it was finally built. It is a steel frame. Eric realises that what he had in mind for Nick to reuse is an earlier version that had a steel part and a concrete part. He is not sure if this earlier version is documented somewhere in the archive. Rather than go through the paper archive again, Eric simply sketches the design for Nick. Eric's sketch also shows the load path concept much more clearly than the CAD drawing would have done, which helps Nick to understand the design. Eric explains to Nick how and why the design evolved. Given the current project they are working on, it would be more appropriate to reuse the earlier composite version. Eric recalls that the specifications of the cooling tower unit itself, which were provided by the HVAC (heating, ventilation and air conditioning) subcontractor, had a large impact on the design. Nick now feels confident enough to design the new cooling tower frame by reusing the same concepts as the Bay Saint Louis cooling tower frame, as well as some of the standard details.

11.5.2 *Observations*

During the above study, the expert's internal knowledge reuse process was observed to be very effective, although it would have been impossible to evaluate his mental retrieval process quantitatively in terms of precision and recall. He was always able to recall directly related past experiences and apply them to the situation at hand. Two key observations in particular characterise the effectiveness of internal knowledge reuse:

- Even though the expert's internal memory was very large (he has over 20 years of experience), he was always able to *find* relevant designs or experiences to reuse.
- For each specific design or part of a design he was reusing, he was able to retrieve a lot of contextual knowledge. This helped him to *understand* this design and apply it to the situation at hand. When describing contextual knowledge to the novice, the expert explored two contextual dimensions: the *project context* and the *evolution history* (Figure 11.5).

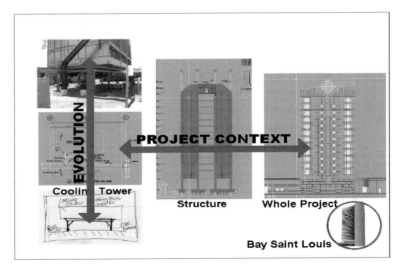

Figure 11.5   Dimensions of knowledge understanding in the process of knowledge reuse

The *observed process* of *internal knowledge reuse* is formalised into three steps:

(1)   finding a reusable item
(2)   exploring its project context in order to understand it and assess its reusability
(3)   exploring its evolution history in order to understand it and assess its reusability.

These observations of internal knowledge reuse are the basis for supporting external knowledge reuse from a corporate memory.

11.5.3   Needs, interaction metaphors and prototypes

Design as reflection-in-action

CoMem™ (Corporate Memory) is the latest in a line of research projects on design KM conducted at the Project-Based Learning Lab at Stanford University. These research projects are based on Schön's reflective practitioner paradigm of design (Schön, 1983). Schön argues that every design task is unique, and that the basic problem for designers is to determine how to approach such a single unique task. Schön places this tackling of unique tasks at the centre of design practice, a notion he terms *knowing-in-action:*

> 'Once we put aside the model of Technical Rationality which leads us to think of intelligent practice as an application of knowledge . . . there is nothing strange about the idea that a kind of knowing is inherent in intelligent action . . . it does not stretch common sense very much to say that the know-how is in the action – that a tight-rope walker's know-how,

for example, lies in and is revealed by, the way he takes his trip across the wire .... There is nothing in common sense to make us say that the know-how consists in rules or plans which we entertain in the mind prior to action.'

(Schön, 1983, p. 50).

To Schön, design, like tightrope walking, is an *action-oriented* activity. However, when knowing-in-action breaks down, the designer may consciously transition to acts of reflection. Schön calls this *reflection-in-action*. In a cycle which Schön refers to as a *reflective conversation with the situation*, designers reflect by *naming* the relevant factors, *framing* the problem in a certain way, making *moves* toward a solution and *evaluating* those moves. Schön argues that whereas action-oriented knowledge is often tacit and difficult to express or convey, what *can* be captured is reflection-in-action.

Semantic modelling engine

This reflection-in-action cycle forms the conceptual basis of knowledge capture in the *semantic modelling engine* (SME) (Fruchter, 1996). SME is a framework that enables designers to map objects from a shared product model to multiple semantic representations and to other shared project knowledge (Figure 11.6). In SME, a *project object* encapsulates multiple *discipline objects*, and a discipline object encapsulates multiple *component objects*. Each SME object can be linked to graphic objects from the shared 3D product model, or to other shared project documents or data (such as vendor information, analysis models, sketches or calculations).

SME supports Schön's reflection-in-action by enabling the designer to declare his/her particular perspective on the design (i.e. *framing* the problem) by creating a discipline object. Next he/she proceeds to *name* the individual components of the problem as he/she sees it by creating component objects. SME discipline objects are exported to external analysis tools to derive building behaviour and evaluate it by comparing it to functional requirements. The designer uses these as the basis for making design decisions, i.e. making *moves* towards the solution and *evaluating* those moves.

SME acts as an integration environment to support the development of a shared building model that uses an AutoCAD graphic representation as the central interface among designers (human-to-human) and as the gateway to tools/services (human-to-machine) in support of cross-disciplinary design. It enables designers to share and explore designs, capture multi-criteria semantics, design rationale critiques, explanations and change notifications.

SME enables a design-build team to explore the different cross-disciplinary issues (Fruchter, 1996) and to:

- *Augment* shared graphic product models with (1) the team members' intents, interests and responsibilities, and (2) formal design rationale

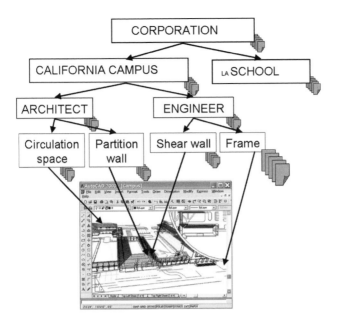

Figure 11.6   Project example using the semantic modelling engine

(e.g. links to online design calculations, vendor product information, financial documents, etc.) and informal design rationale, knowledge and information (e.g. RECALL[TM] sessions).

- *Gather networked information* by using the discipline sub-systems to customise their search for additional discipline information.
- *Analyse* and *evaluate* the discipline sub-systems to derive the building's behaviour and compare it to functional requirements.
- *Explain* the results to other members of the team.
- *Capture* design perspectives at different levels of granularity, i.e. the component, discipline and project levels.
- *Infer* shared interests and route change notifications with regard to a modified component or discipline sub-system.
- *Search* for relevant information based on keywords or graphic object selections in the shared building model.

The information and knowledge related to the shared product model are organised as follows:

- *Graphic objects* contain Drawing Interchange File (DXF) representations of the graphic model entities.
- *Discipline objects* encapsulate features for a particular perspective. A discipline object has two primary attributes: a list of *component classes* and a list of *component objects*. Component classes provide an ontology to describe the semantic meaning of the graphics within a context. This

ontology can be defined or augmented by the user at run-time. The list of component objects is edited by the user and contains the instances from a particular graphic model that are relevant to an interpretation.

- Component objects capture the link between graphic entities and symbolic entities. We define a component to be a constituent element of a design that has meaning to a designer within a particular context. The basic elements of a component object are a component class, an identifier or *component name* and a list of graphics objects. Other information objects can be linked to component objects such as *note objects, W-Doc objects (hyperlinked objects)* and *notification objects.* Component objects allow graphic entities to have multiple meanings within different discipline objects. For example, the same physical object can be named as 'shear wall' by the structural engineer and as 'partition wall' by the architect.

- *Person objects* serve as a record of the project participants and their declared roles and interests. A person object consists of the designer's name, a user-name, a user-password, an email address, a list of responsibilities and a list of interests. Person objects can be added, updated and deleted by the users. The lists of interests and responsibilities are used to infer which team members should be sent email notifications about changes to a portion of the design.

- *Note objects* contain text written by the project members. Note objects are used to capture the design rationale or other design-related information that a designer traditionally records in notebooks, memos, etc. Notes are encapsulated in component objects to describe design requirements or intents.

- *W-Doc objects* provide a mechanism for linking a component object to sources of information. References to any information published on the world-wide web (WWW) can be handled. A component in the graphic model can be linked to component specification sheets, code pages, structural details, schedule and cost information, vendor catalogue information, services available on the WWW or a photograph of the construction site.

- *Change notification objects.* SME facilitates communication among designers using change notification objects. These notifications are used to solicit feedback, give approval, broadcast changes or initiate negotiations. SME routes change notifications automatically based on multiple interpretations of graphic objects and interests expressed by the design team members.

The ProMem™ (Project Memory) system (Fruchter *et al.*, 1998; Reiner and Fruchter, 2000) takes the SME as its point of departure and adds to it the time dimension. ProMem™ captures the evolution of the project at the three levels of granularity identified by SME as emulating the structure of project knowledge: project, discipline and component. ProMem™ automatically versions each SME object every time a change is made in the design or additional knowledge is created. This versioning is transparent to the designer. The designer is able to go back and flag any version to indicate its *level of importance* (low, conflict or milestone) and its *level of sharing* (private, public or consensus). The ProMem™ system architecture consists of three modules: the *shared graphic modelling environment*, i.e. AutoCAD, *VisionManager*, a module that was developed for KM and integrated within AutoCAD to provide the link between the shared graphic model and the third module, i.e. a commercial *object-relational database management system* (ORDBMS) for storing the evolution of the product models. The system was implemented using C, AutoLisp, DCL and SQL statements. It runs on Sun workstations and Windows platforms. It has been tested and validated in a dozen building projects in both education and industry test case projects, as well as in other vertical markets such as automotive, manufacturing and aerospace industry pilot product development projects (Fruchter *et al.*, 1998).

CoMem™ extends ProMem™ first by grouping the accumulated set of project memories into a corporate memory, and second by supporting the designer in reusing design knowledge from this corporate memory in new design projects. This support for knowledge reuse is based on observations of internal knowledge reuse by designers at work. This knowledge reuse is not limited to designed components and sub-components, but includes the evolution, rationale and domain expertise that contributed to these designs. Here we echo Schön's contention that design expertise lies not in 'rules or plans entertained in the mind prior to action' but in the action itself.

The CoMem™ human–computer interaction experience is based on the principle of *'overview first, zoom and filter, and then details-on-demand'* (Shneiderman, 1999). This principle is used to design a user experience that is based on the empirical observations of internal knowledge reuse.

The objective of CoMem™ is to enable the user to interact with a potentially huge archive of design knowledge from previous projects. Specifically, CoMem™ must support the three activities in the reuse process: *find, explore project context*, and *explore evolution history*. CoMem™ comprises three modules that support each of those three activities: overview, project context explorer and evolution history explorer.

The CoMem™ *overview* supports the designer in finding reusable items. Assuming that the designer does not know *a priori* where in the corporate memory reusable items can be found, the overview should initially show all items. The overview needs to provide a succinct 'at a glance' view of the entire corporate memory. CoMem™ uses a map metaphor for the overview.

The hierarchy consisting of projects, disciplines and components is visualised as a series of nested rectangles using the squarified treemap (Bruls *et al.*, 1999) technique (Figure 11.7). The area on the map allocated to each item is based on a measure of how much knowledge this item encapsulates, i.e. how richly annotated it is, how many times it is versioned, how much external data are linked to it. Each item on the map is colour-coded by a measure of relevance to the designer's current task, and so the overview

Figure 11.7   Overview module in CoMem™ user interface

gives the designer an indication of which 'regions' of the corporate memory contain potentially reusable items. This relevance measure is based on an information retrieval technique called latent semantic analysis (Landauer and Dumais, 1995) of the textual data in the corporate memory. CoMem™ allows the user to filter out items from the overview using dynamic querying. In a dynamic querying environment, search results are instantly updated as the user adjusts sliders or selects buttons to query a database (Shneiderman, 1994). The designer can filter based on relevance, date, person or keyword. Items that are filtered out can either appear greyed out or are not drawn at all, leaving more space for the remaining items.

Once the user has selected an item from the overview, the CoMem™ *project context explorer* supports the designer in exploring this item's project context. This shows the project and discipline to which this item belongs, as well as related components, disciplines and projects. The item selected from the overview becomes the focal point of the project context explorer. CoMem™ uses a fisheye lens metaphor for the project context explorer. A fisheye lens balances local detail with global context (Figure 11.8).

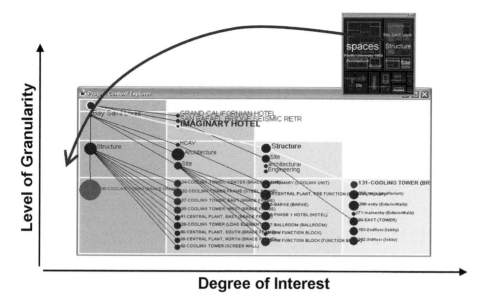

Figure 11.8   Project context explorer module in CoMem™

Given a user-specified focal point, CoMem™ uses the fisheye formulation (Furnas, 1981) to assign a degree of interest to every item in the corporate memory. Items with a higher degree of interest are displayed more prominently in the project context explorer. In the CoMem™ project context explorer, each object is positioned in the vertical axis according to its level of granularity, and in the horizontal axis according to its degree of interest. This visualisation emphasises structural relationships in the hierarchy that are obscured in the map, and facilitates effective exploration of the focal node's context.

Finally, in the CoMem™ *evolution history explorer*, the designer can explore the evolution history of any item selected from the overview. This view tells the story of how this item evolved from an abstract idea to a fully designed and detailed physical artifact or component. CoMem™ uses a story-telling metaphor for the evolution history explorer (Figure 11.9). This is based on our observation that the most striking means of transmitting knowledge from experts to novices in AEC design offices is through the recounting of experiences from past projects. Stories convey great amounts of knowledge and information in relatively few words (Gershon and Page, 2001).

Each version is represented by a circle colour-coded to indicate this version's level of importance (low, conflict or milestone) and level of sharing (private, public or consensus). The original designer working on the specific version of the project provides these levels. Next to this circle, thumbnails appear for any CAD objects, sketches, documents or notes linked to this version. The designer is able to click on any of the thumbnails for a larger

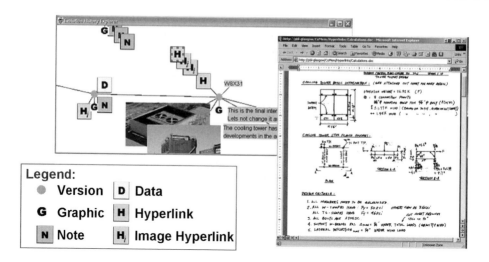

Figure 11.9   Evolution history explorer module in CoMem™

view of this content. The designer is able to filter out versions based on the levels of importance and sharing (Figure 11.9).

CoMem™ can be a valuable vehicle in alternative exploration and decision support for knowledge reuse from past projects. We have integrated the CoMem™ prototype with an innovative interactive room (iRoom) environment that controls multiple computers linked to large displays (e.g. SmartBoards or projection screens) developed by the human–computer interaction (HCI) group in the Computer Science Department at Stanford (Winograd 1999, unpublished working paper 'Towards a Human-Centered Interaction Architecture', http://graphics.stanford.edu/projects/iwork/papers/humcent/index.html). The CoMem™-iRoom can facilitate effective exploration and comparative studies of rich content in context during project team meetings (Figure 11.10). Chris Michaelis of Intel in Arizona deployed the integrated CoMem™-iRoom prototype demo in February 2003 to explore its potential use with project managers, project team members and sub-contractors at Intel. He says:

'CoMem™ provides great benefits by representing project information and knowledge in various forms for each user type to gain the greatest benefit. Having the actual context that the information was created in stored with the information makes reuse so much more powerful. Project managers can use the CoMem™ Overview to find related components from the entire corporate memory to help solve (and avoid) project issues while taking that same information in the "storyteller" Evolution History Explorer metaphor to work with subs to determine the reuse capabilities. The Evolution History Explorer allows for more than just the end state to be reusable, it allows users to examine all the progressive states with the

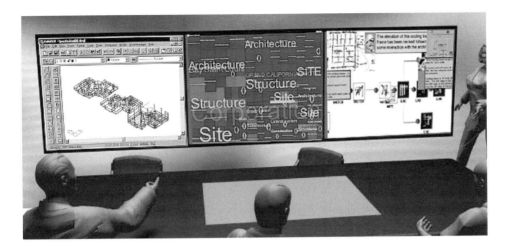

Figure 11.10   CoMem^{TM} i-Room prototype

complete context of the knowledge creation making for many reusable knowledge components. Very powerful in its presentation of complete knowledge within context!'

11.6   Conclusions

Reusing designs and design knowledge from an external repository of knowledge from previous projects is an important process that often fails. We attribute this failure to the fact that state-of-practice archiving systems (such as archives of paper drawings or electronic files arranged in folders) do not support the designer in *finding* reusable items and *understanding* these items in context in order to be able to reuse them. We argue that the designer must be able to explore the *project context* and *evolution history* of an item in order to understand it and reuse it. We present prototype systems that support knowledge capture, sharing, finding and reuse through project context exploration and evolution history exploration in a large corporate memory that consists of informal tacit and formal explicit knowledge.

RECALL^{TM}, ProMem^{TM} and CoMem^{TM} support interactions that are radically different from those found in current practice (e.g. scrolling through project directories and folders using Windows Explorer used today). They use metaphors to link these interactions to familiar experiences or objects. CoMem^{TM} captures knowledge and information in rich contextual informal media, i.e. sketch, video and audio. CoMem^{TM} supports finding using the Overview that employs a map metaphor. Once a potentially reusable item is found, the designer can explore its project context using the Project Context Explorer which employs a fisheye lens metaphor. The designer can explore the item's evolution history using the evolution history explorer which employs a story-telling metaphor and the designer can asses the reusability of the items by understanding the context in which they were created in ProMem™.

Ongoing work is now focusing on usability tests and evaluating the impact of the new interaction metaphors in the decision process of an AEC design–build project.

References

Bhatta, S., Goel, A. and Prabhakar, S. (1994) Innovation in analogical design: a model-based approach. In: *Proceedings of the 3rd International Conference on Artificial Intelligence in Design (AID-94)*, August, Lausanne, pp. 57–74.

Blomberg, J., Giacomi, J., Mosher, A. and Swenton-Wall, P. (1993) Ethnographic field methods and their relation to design. In: *Participatory Design: Principles and Practices* (D. Schuler, ed.), pp. 123–55. Lawrence Erlbaum Associates, New Jersey.

Bruls, D.M., Huizing, C. and van Wijk, J.J. (1999) Squarified treemaps. In: *Data Visualization 2000, Proceedings of the Joint Eurographics and IEEE TCVG Symposium on Visualization, 2000* (W. de Leeuw and R. van Liere, eds), pp. 33–42. Springer, Vienna.

Cross, N. and Roozenburg, N. (1992) Modelling the design process in engineering and architecture. *Journal of Engineering Design*, **3** (4), 325–37.

Cross, N., Christiaans, H. and Dorst, K. (1996) *Analyzing Design Activity*. John Wiley and Sons, Chichester.

Davenport, T. and Prusak, L. (1998) *Working Knowledge: How Organizations Manage What They Know*. Harvard Business School Press, Boston.

Domeshek, E. & Kolodner, J. (1993) Finding the points of large cases. *Artificial Intelligence for Engineering Design, Analysis and Manufacturing*, **7**(2), 87–96.

Eastman, C.M. (1996) Cognitive processes and ill-defined problems: a case study from design. In: *Proceedings of an International Joint Conference on Artificial Intelligence*, May, Washington DC, pp. 675–99.

Finger, S. (1998) Design reuse and design research. In: *Proceedings of Engineering Design Conference on Design Reuse*, June, Brunel University, Professional Engineering Publishing Ltd, pp. 3–10.

Fruchter, R. (1996) Conceptual, collaborative building design through shared graphics. *AI in Civil and Structural Engineering*, **11**(3), 33–41.

Fruchter, R. (1999) Architecture/engineering/construction teamwork: a collaborative design and learning space. *Journal of Computing in Civil Engineering*, **13**(4), 261–70.

Fruchter, R. (2003a) Innovation in engaging learning and global teamwork experiences. In: *Proceedings of ASCE Computing in Civil Engineering Conference* (I. Flood, ed.), Nashville, Nov.

Fruchter, R. (2003b) Degrees of engagement in interactive workspaces. In: *Proceedings of SID2003 2nd Social Intelligence Design Workshop* (D. Rosenberg, T. Nishida and R. Fruchter, eds), London, July.

Fruchter, R. and Demian, P. (2002) CoMem: designing an interaction experience for reuse of rich contextual information from a corporate memory. *AIEDAM*

International Special Issue on 'Human Computer Interaction in Engineering Context' (I. Parmee and I. Smith, eds), **16**, 127–47.

Fruchter, R. and Yen, S. (2000) RECALL in action. In: *Proceedings of the 8th International Conference (ICCCBE-VIII) Computing in Civil and Building Engineering* (R. Fruchter, K. Roddis and F. Pena-Mora, eds), 14–16 Aug, Stanford, California.

Fruchter, R., Reiner, K., Leifer, L. and Toye, G. (1998) VisionManager: a computer environment for design evolution capture. *CERA: Concurrent Engineering: Research and Application Journal*, **6**(1), 71–84.

Furnas, G.W. (1981) The FISHEYE view: a new look at structured files. In: *Readings in Information Visualization* (S.K. Card *et al.*, eds). Morgan Kaufmann, Los Altos, California, pp. 312–30.

Gershon, N. and Page, W. (2001) What storytelling can do for information visualization. *Communications of the ACM*, **44**(8), 31–7.

Goel, V. (1995) *Sketches of Thought*. MIT Press, Boston.

Goldschmidt, G. (1991) The dialectics of sketching. *Creativity Research Journal*, **4**(2), 123–43.

Hu, X., Pang, J., Pang, Y., Atwood, M., Sun, W. and Regli, W.C. (2000) A survey on design rationale: representation, capture and retrieval. In: *Proceedings of ASME Design Engineering Technical Conference, 5th Design for Manufacturing Conference* (DETC 2000/DFM-14008), 10–13 Sept, Baltimore, Maryland.

Kosslyn, S. (1981) The medium and the message in mental imagery: a theory. *Psychological Review*, **88**, 46–66.

Kuffner, T.A. and Ullman, D.G. (1990) The information requests of mechanical design engineers. In: *Proceedings of 2nd International Conference on Design Theory and Methodology*, Sept, Chicago, pp. 167–74.

Landauer, T.K. and Dumais, S.T. (1995) A solution to Plato's problem: the latent semantic analysis theory of acquisition, induction and representation of knowledge. *Psychological Review*, **104**, 211–40.

Mabogunje, A. and Leifer, L. (1997) Noun phrases as surrogates for measuring early phases of the mechanical design process. In: *Proceedings of DETC'97*, ASME, Sacramento, California.

Maher, M.L. (1997) CASECAD and CADSYN: implementing case retrieval and case adaptation. In: *Issues and Applications of Case-Based Reasoning in Design* (M. L. Maher and P. Pu, eds). Lawrence Erlbaum, Mahwah, New Jersey.

National Materials Advisory Board (1991) *Enabling Technologies for Unified Life-Cycle Engineering of Structural Components*. Publication NMAB-455. Committee on Enabling Technologies for Unified Life-Cycle Engineering of Structural Components, National Materials Advisory Board, Washington, DC.

Nelson, T. (1990) The right way to think about software design. In: *The Art of Human–Computer Interface Design* (B. Laurel, ed.). Addison-Wesley, New York.

Nonaka, I. and Takeuchi, H. (1995) *The Knowledge-Creating Company: How Japanese Companies Create the Dynamics of Innovation*. Oxford University Press, Oxford.

Olszweski, E.J. (1981) *The Draughtsman's Eye: Late Renaissance Schools and Styles*. Cleveland Museum of Art/Indiana University, Cleveland, Ohio.

Reiner, K. and Fruchter, R. (2000) Project memory capture in globally distributed facility design. In: *Proceedings of the 8th International Conference (ICCCBE-VIII)*

Computing in Civil and Building Engineering, Stanford, California, 14–16 Aug, pp. 820–27.

Rosson, M.B. and Carroll, J.M. (2001) *Usability Engineering: Scenario-Based Development of Human Computer Interaction.* Morgan Kaufmann, Los Altos, California.

Schön, D.A. (1983) *The Reflective Practitioner.* Basic Books, New York.

Shneiderman, B. (1994) Dynamic queries for visual information seeking. In: *Readings in Information Visualization* (S. K. Card *et al.*, eds). Morgan Kaufmann, Los Altos, California, pp. 236–43.

Shneiderman, B. (1999) Supporting creativity with advanced information-abundant user interfaces. In: *Human-Centred Computing, Online Communities, and Virtual Environments* (R. Earnshaw *et al.*, eds), pp. 469–80. Springer, London.

Stiedel, R.F. and Henderson, J.M. (1983) *The Graphic Languages of Engineering.* Wiley, New York.

Tversky, B. (1999) What does drawing reveal about thinking? In: *Proceedings of Visual and Spatial Reasoning in Design* (J.S. Gero and B. Tversky, eds), 15–17 June, MIT, Cambridge, Massachusetts.

Winner, R.I., Pennell, J.P., Bertrend, H.E. and Slusarczuk, M.M.G. (1988) *The Role of Concurrent Engineering in Weapons System Acquisition.* IDA Report R-338. Institute for Defense Analysis, Alexandria, Vancouver.

12 Building a Knowledge-Sharing Culture in Construction Project Teams

Patrick S.W. Fong

12.1 Introduction

To enhance competitiveness and meet their goals, organisations need to ensure that their employees share their knowledge. Increased sharing of knowledge raises the likelihood of new knowledge being created, tending to support valuable innovation (Nonaka and Takeuchi, 1995). Though organisations can codify some of the knowledge people use, it is easy to find situations that do not fit the codified knowledge of the organisation. This unarticulated knowledge requires communication among people. Orr (1996) finds that photocopier technicians often search for solutions beyond their work manuals. He (1996, p. 2) explains that 'the expertise vital to such contingent and extemporaneous practice cannot be easily codified'. At one point during his study, Orr (1996) finds technicians joking about the usefulness of their manuals. When written documents prove insufficient, people need to access each other's experience to solve more difficult problems. Orr shows how technicians sometimes use narrative to recount each other's previous experience. Technicians might use breakfast or lunch gatherings to share their knowledge. Another account of knowledge sharing demonstrates how employees use informal groups, known as communities of practice, to share what they know (Wenger *et al.*, 2002). In both cases, communication is the key to sharing knowledge.

Knowledge sharing relies on reaching a shared understanding of the underlying knowledge, not just the content but also the context of the knowledge, or 'Ba', to use Nonaka and Konno's (1998) term. Exchanging information represents only a partial view of the knowledge-sharing activity. The essence lies in unveiling and synthesising paradigmatic differences through social interaction.

Knowledge sharing is not restricted to exchanges among the employees of a company. It can occur between employees and customers, or between organisations and firms in entirely different industries (von Hippel, 1988). Some of the very important knowledge types identified in a survey of knowledge-intensive businesses include customer, competitor and product knowledge (Skyrme and Amidon, 1997). The more knowledge is shared about the needs of current and potential customers among project team members, the better they may understand genuine customer requirements. With such knowledge, greater value may be created for customers because the company can ensure that its end products better satisfy customer needs and requirements. Accordingly, the products might have a better chance of success in the market place. In the same vein, shared competitor knowledge could be helpful in developing products ahead of market requirements (getting products to market ahead of competitors or developing products on schedule). It could yield high value to customers (extending any product's success in the marketplace), and possibly improve product performance (better overall product performance than that of competitors). In addition, shared product knowledge (product advantage, disadvantage, strengths, history and technologies) may be important in order to improve development productivity (reducing development costs) and production costs (reducing overall production costs).

It is clear that sharing diverse knowledge can enhance problem solving, as well as creating the culture required for knowledge creation. Again, communication is the key to knowledge sharing.

This chapter is divided into four sections. Following the Introduction, Section 12.2 presents an industrial case study of knowledge sharing involving construction professionals and the client's specialist consultants. The importance of socialising, sharing positive as well as negative experiences, and knowledge 'shielding' to protect a client from its competitors are discussed. Section 12.3 is a discussion of the impact of competition, communication and different knowledge domains on knowledge sharing. The importance of openness, motivation, trust and time are highlighted as essential ingredients to facilitate knowledge sharing. Concluding remarks based on the findings from the case study are made in Section 12.4.

12.2   Case study

12.2.1   Project team background

The members of the residential development project team under study were all experienced construction professionals. Their professional disciplines encompassed project management, interior design, architecture, building services engineering, structural engineering, lands consulting, environment consulting and landscape architecture. The client also engaged specialist consultants whose expertise could be beneficial to this project. These included interior designers and a landscape architect. It was not uncommon on projects of a smaller scale for the architects to be responsible for designing landscaping as well as interior design. In addition, the client also appointed a property consultant specialising in land matters to oversee lease interpretation, lease modification, etc. For project management services, the client used its in-house project management department. This department played an active role in the development of the project, by attending most of the meetings, constantly monitoring the project's progress and having direct discussions with government officials on various issues affecting its development.

12.2.2 Socialising with other project team members

In the residential project, the opportunities for team members to socialise with each other appeared limited. This was probably due to the fact that some team members, like the interior designer, had been appointed to the project only recently. The structural engineer, hitherto uninvolved in the project, had suddenly found himself appointed to it upon his predecessor's departure. When the senior project manager was interviewed about the team's social activities, he started to realise that not enough had been done. He realised that the fast pace of development was putting enormous time pressures on all the consultants, who were working simultaneously on several projects. This caused people to become very focused on their tasks and to neglect the fostering of relations within the team. Some suggested that social events would help improve team interaction, so that knowledge could be exchanged more naturally. They believed that social events could enhance relationships, decreasing the distance between team members. This in turn would lead to improved knowledge sharing.

> 'Socialising with other project team members is not common and usually not done intentionally – mostly people have lunch together. Social activities are good for the success of projects, since interpersonal relationships are very important.' (Associate Architectural Director)

He felt that informal conversation could help one to understand issues not previously dealt with. The interior designer of the residential clubhouse also thought it important to maintain social contact, developing relations on a more relaxed level.

> '. . . You establish a kind of rapport. . . . We are all in the same boat. It's nice to talk about something else, to find out more about the person than just knowing them purely on a professional level.'

He saw socialising beyond the professional working relationship as very important in moving a project forward. Some team members found that during social events, they would talk about work-related issues in a relaxed fashion. Others preferred not to discuss the project, using the occasion purely to get to know one another. Some team members suggested that while socialising, they sometimes developed unexpected insights or solutions from the more informal discussions. The building services engineer echoed others in saying that when an issue was aired during a social event, everyone was generally more willing to exchange views. He suggested organising social activities where team members would avoid talking about the project, using these occasions purely to build relationships.

The senior project manager found that using the telephone to contact other team members was more personal than email. Telephone conversations tended to be more interactive, dynamic and instant than typing out email messages, which took time and seemed more clinical.

12.2.3   Sharing positive as well as negative experiences

From the outset, all project team members had been expected to freely share their knowledge and expertise, as this appeared to be the norm in professional appointments. This willingness to behave co-operatively and integrate individual knowledge with that of other team members is fundamental to the process of knowledge creation (Nonaka, 1994). The assistant project manager confessed that there was personal bias as to what was important and should be shared.

The senior project manager found that team meetings were a good platform upon which to trade knowledge. He added that one characteristic of construction projects was that multiple meetings were necessary, as each profession's work was affected by everybody else involved in the project. He observed that generally it could be awkward for people to reveal bad experiences, but that within the project team, because of its established trust, people found ways to unburden themselves of their errors. He found that the act of sharing knowledge was not difficult – but that finding time was a key issue.

The associate architectural director confessed that no one person could solve all the problems encountered in any given career. People need to learn and exchange ideas in the office, even with their junior staff and with the project team. He found that team members often referred in meetings to other projects from which they had gained either personal experience or insights from friends. The client might also reveal information from previous projects that would usually be precise and relevant to other team members.

The team members in this project saw personal experience as critical in their offering of professional services to clients. The client's organisation, especially the project management department, had a very open culture of sharing knowledge and experience. They revealed their successes as well as their failures, reasoning that no single team member could have had full exposure to all the different types of project situations.

> 'In our in-house debriefing sessions, we share mistakes as well as successes.' (Assistant Project Manager)

It appeared, partly due to the open culture of the client organisation and partly due to the already established knowledge-sharing culture within the consulting firms, that the team members generally traded both their positive and their negative experiences. The failures served to remind the team members not to pursue the same paths as they had followed previously. The successes acted as models of reference for others to follow.

'I can see how the knowledge from one project can be quickly applied to another project. This is due to shared project experience.' (Associate Architectural Director)

Knowledge shared and experience gained by a team member on one project enhanced overall learning. As some team members suggested, this was extremely important in construction consulting, in which not all team members had been exposed to the same depth of knowledge and experience from projects undertaken previously.

12.2.4 Shielding knowledge sharing

Due to the confidential nature of the design information and the special project features, the consultants were asked not to have the same in-house teams working on key competitors' projects. Accordingly, consultants' in-house teams were advised not to engage in direct communication with other in-house teams working for different clients, in case confidential information was leaked. Both the architects and the interior designers were specifically reminded of this. The main reason for the client's concern was that there were distinctive features of the design that would make the product sell. Leaking those features could mean losing the competitive advantage. The client's fear was that if confidential information about this project were leaked, the competitors would possess knowledge that would distinctly disadvantage the client organisation. Competitors could adopt the special design features, apply them to their projects and then launch these hijacked developments ahead of time.

The architect and interior designer explained the system used to avoid the spread of knowledge between different in-house teams.

'We have different teams in the office working on other residential projects and the current project. Some developers prefer to keep their developments secret. We respect their privacy and thus may not use ideas from that project in their competitors' projects.' (Associate Architectural Director)

'I only work on this client's projects or on joint-venture projects with other developers. Other teams work on other developers' projects.' (Associate Interior Design Director)

'Some project managers will remind us not to have the same team working on arch competitors' projects due to the extreme sensitivity of such competition. We also try to avoid communication with other in-house teams that work for different developers.' (Associate Interior Design Director)

In addition, the clubhouse interior designer suggested that there was secrecy in their profession about the projects they were currently working on. They usually did not let interior designers from other firms know what projects they were working on until the project had already started on site. The two interior designers working on the tower blocks made similar comments. They all stressed that they did not contact other professionals from the same specialisation for knowledge or advice. They would rather rely on professionals within other disciplines, or on suppliers or contractors.

'No, what we don't want to do in the early stages of the project is let anyone know that we're working on it. Even though we've signed a contract and have paid a retainer, if word gets around that there's work available on a particular project, then other companies might try to take the deal away. So in the initial stages, yes, we very much have to rely on our in-house knowledge and sources.' (Clubhouse Interior Designer)

'We try to avoid telling the suppliers, because they might mention us in conversation to someone. Normally, most design firms are very shrewd and they'll say we have a project but they won't specify what project exactly. If they have to write to a supplier requesting something, they will just say something like "clubhouse" project or "hotel" project – something open-ended, giving no clues. We're very aware of it. It's always been the case since I first started working in interior design, that people are quite protective about their project list. We don't discuss anything until we're actually going on site. We're then free to talk about it to suppliers.' (Clubhouse Interior Designer)

The interior designers for the tower blocks likewise stressed that they seldom contacted colleagues in the same profession. The clubhouse interior designer agreed that the industry was highly competitive and that, accordingly, there was a profound lack of trust.

12.2.5 Contributory factors

Openness

The project managers consistently delivered the message that communication was vital and stressed the importance of knowledge sharing, encouraging team members to adopt this open culture. Some team members described themselves as having very open personalities, not shielding any knowledge. Team members indicated that there was no lack of trust, nor was there a secretive culture. A successfully integrated design relied on all team members, who were transitory, coming from different organisations purely to get the project completed. There was no reason to withhold their knowledge and experience, as they were not in competition with each other.

> 'From what I've seen so far, there's been no lack of trust or any secrecy between the individual consultants. Everyone has his own area of expertise and is quite willing to talk about it to solve problems – the way to produce a better result, by taking a different approach.' (Clubhouse Interior Designer)

> 'It is different from university life, where students will hide some secret weapons. In the reality of this industry, people are willing to share. Through sharing, others may comment on your ideas and you will learn even more. There is no secret about knowledge; it is different experience and exposure.' (Associate Architectural Director)

> 'Sharing knowledge is kind of self-initiating. It very much depends on personal character. Some people like to talk to people and some don't.' (Architectural Director)

The remarks from various team members generally reflected a willingness to communicate and interact with one another, to exchange knowledge and experience.

The formal and informal meetings served as channels for team members to discuss issues and problems openly, with the prime objectives of resolving them expeditiously, to get the project going. Due to the very tight programme schedule, team members could not afford to dwell on any single issue for too long. The clubhouse interior designer therefore found that informal meetings generally facilitated open discussion.

> 'I have to say that informal meetings are probably more productive.... Formal meetings have to be minuted and nobody wants to be seen saying something which could later backfire. Informal meetings are where you can get down to the bones, and with not being minuted, you get a much better idea of what people are looking for.'

During an informal meeting, the building plans for the top two floors in each residential tower were discussed. These were more luxurious than the other units. The flats were around 120 m^2. Team members started to discuss the size and layout of the living room and bedrooms, wondering about how to maintain a 4 m distance between the television and the sitting areas. Parallels were drawn with a previous project. The architectural director raised the issue of 'feng shui', with some bedroom doors facing the main entrance to the flat, which meant that wealth and luck could escape from the unit. The senior project manager saw this as a personal preference but complimented the architect for sharing it openly. In conclusion, team members were found to share their design knowledge, experience and ideas openly. This was important, given the breadth of experience possessed by the team as a whole.

Motivation

As in many multi-disciplinary situations, team members were not rewarded for sharing their knowledge or experiences, as this was merely part of their professional duty. The following quotes by various team members highlight their motivation for exchanging knowledge. The simple reason, as reinforced by an interior designer, was that the nature of project work requires collective wisdom and teamwork.

'We need to share knowledge, as we cannot deal with the project alone. We need teamwork.'

'Project work requires the collective wisdom of the team, unlike fine art, which is an individual masterpiece.'

The architectural director observed that the motivation to share knowledge stems purely from personal interest.

'Motivation to share knowledge is driven by personal interest.'

The clubhouse interior designer saw helping others as a gratifying experience.

'It's nice to be able to help someone out. If they have a query or a problem that they can't solve and they come to you, and you've had that problem before and you can solve it, it's very gratifying to know that you can rely on each other to solve problems.'

Overall, team members seemed highly motivated to share their expert knowledge and experience with others during the design process.

Time pressure

All the current consultants indicated that they needed to strictly account for the time they spent on projects – typical of the consultancy sector in general. Hence, within the project's fixed time frame established by the respective companies, consultants tended to have the least time in which to maximise profits. Some team members indicated that this was a common measure that companies used to gauge their employees' performance. This approach was considered to potentially constrain both individual and collective knowledge bases, possibly stifling creativity. In such an environment, it was in the consultant's interests to complete project work successfully and quickly, with little incentive to contribute towards creative solutions.

The time pressures experienced by team members tended to be largely imposed by the client and their companies, and were considered to impose constraints on knowledge creation. Instead of experimenting with or exploring new ideas, team members might adopt existing solutions to new problems.

The associate architectural director saw time pressure as two sides of the same coin, facilitating as well as inhibiting knowledge sharing. He found that people shared when they were pressed to resolve an issue, but, equally, they might not share when lacking time.

> 'One thing that facilitates knowledge sharing is a deadline to resolve an issue.' (Associate Architectural Director)

> 'The major barrier to sharing knowledge is lack of time. We already expend all our time on projects.' (Associate Architectural Director)

Team members were asked specifically about barriers to knowledge sharing. They all cited lack of time as the only major issue, since the design programme schedule was very pressing and team members could not dwell on a single issue or problem for very long. A culture of secrecy or a lack of trust did not appear to be problems in the knowledge-sharing process among multi-disciplinary project team members.

12.3   Discussion

The following discussion illustrates three aspects of knowledge sharing: competition, communication thickness and knowledge sharing from different knowledge domains. In discussing each aspect, the actions of the project team are highlighted and compared with findings in the current literature. Lastly, influences on the knowledge-sharing process are identified and explained.

12.3.1   Competition and knowledge sharing

The residential development project illustrates that knowledge sharing can be a double-edged sword in attempts to foster competitive advantage. When competitive advantage partially depends upon the non-imitability of knowledge used in product strategies, and when knowledge sharing comes at the cost of increased knowledge leakage to competitors, thereby facilitating imitation, the company's competitive position could well be eroded rather than improved. It is tacit – not explicit – knowledge that most accurately fits the description of resources to which sustainable competitive advantage and associated returns confer (Leonard and Sensiper, 1998). In contrast to easily traded and widely accessible resources, idiosyncratic and scarce knowledge qualifies as a strategically significant resource (Winter, 1987).

In this project, sharing in-house knowledge among the architects or interior designers could yield potential losses through unwanted knowledge diffusion, should company designs be imitated elsewhere. This could result in restrictions being considered on knowledge sharing among in-house members. Knowledge once articulated and codified may be easier and less costly to replicate and leverage internally. Simultaneously, however, codification and easy access to shared knowledge increases the risk of imitation outside the company. Consequently, companies have to weigh the potential for, and the speed of, internal replication against the threat of expropriation through imitation (Kogut and Zander, 1992).

In Hong Kong, residential development projects are more susceptible to competition, more sensitive to customer needs and more vulnerable to market forces as each year tens of thousands of residential units, both new and second-hand, move onto the market. Acutely aware of this market vulnerability, the client's organisation had a further security risk to consider. The heavy competition in Hong Kong's construction industry results in energetic bidding for projects by professional services firms. They cannot afford to serve one particular client alone, as the returns would not be sufficient. Accordingly, they often end up working for clients who may be in competition with each other. This was very much the case in this residential project, with both the architectural and interior design firms working for the client company as well as for its competitors. The project managers explicitly requested that the designers not have the same teams serving both the project in question and those of any rivals. In this way, the risk of confidential information and design details being leaked to competitors could be minimised. They were further advised to have totally separate in-house teams providing professional services to the client and to their competitors, with absolutely no direct communication between them. Teams were simply not allowed to share knowledge. At the same time, the sharing of information and insights across teams and professions would appear to be important in stimulating knowledge creation.

This would seem to run contrary to the knowledge-sharing principle, whereby employees are encouraged to share their knowledge. All team members from the architectural and interior design firms were fully aware of this constraint. They seemed to appreciate that their input and discretion were both critical to the success of the product launch. It is interesting to speculate whether such shielding of knowledge may in time erode the working culture within companies, with colleagues communicating less to avoid divulging confidential information.

Furthermore, several interior designers revealed that they would never contact colleagues from other firms to share or seek knowledge. The competition among interior designers was intensely fierce. This lack of knowledge cross-fertilisation might result in some 'reinventing of the wheel'. Negative work experiences might also be needlessly re-enacted, given the potential scarcity of cautionary advice. There is also the risk of self-complacency or stagnation. Design creativity can be triggered through sharing knowledge with professional peers. Interviews with members of other disciplines did not reveal similar levels of knowledge hoarding. It is possible that this tendency could curb knowledge advancement in the interior design industry.

12.3.2 *Communication thickness and knowledge sharing*

Because of their specialist professional knowledge, team members had to share both tacit and explicit knowledge in order to meet a general deadline most effectively. The process of creating knowledge in teams consistently begins with communication between individuals.

This study revealed a connection between the medium of communication and the type of knowledge shared. Thicker communication was associated with the sharing of tacit knowledge among team members in the project. This result is consistent with the critical social theory (Habermas, 1987), which suggests that thicker communication media are preferred for sharing more complex information and knowledge. Most personal experience or personal tacit knowledge is shared through the active interpretation of the sender and receiver. This is in line with the findings of Nonaka and Takeuchi (1995), who suggest that Japanese corporations have invested in expensive corporate retreats to withdraw to when complex decisions need to be made. In this case, thicker communication enables complicated ideas, opinions, social cues and emotions to be exchanged and interpreted.

Individuals from the project team possessed expertise and knowledge relevant in varying degrees to their project work. This did not guarantee that they would necessarily share their knowledge. Before this could take place, team members needed to recall any relevant information or knowledge, and also be motivated to share it. The negative consequences of failing to share critical knowledge during the design phase could result in sub-standard or faulty design, possibly causing fatalities, injuries or simply mere discomfort. In this project, the failure of the architects to convey their intention of creating a clear span for the roof structure above the swimming pool had resulted in the structural engineer designing the roof using columns, beams and a flat roof structure. This situation was later rectified during an informal meeting and the engineer had to resolve the situation by using a much more expensive solution of a structural steel, lattice roof structure to accommodate the large, unobstructed span. Whatever the scenario, the impact on the project could be considerable and would be best avoided. The more conducive an atmosphere is to knowledge sharing, the more likely it is that problems will be addressed and corrected.

All project team members were located within Hong Kong, enabling them to interact frequently in person through scheduled formal and informal meetings. Ample opportunities were available to share explicit as well as tacit knowledge, allowing hidden knowledge to be exchanged. Personal interaction is regarded as the richest form of communication because it provides multi-mode communication with immediate feedback. Though such interaction provides more opportunity for knowledge sharing, this does not necessarily imply that it will happen naturally and automatically in every team situation. Opportunities for project team members to interact with and observe one another are better mechanisms for transferring tacit knowledge than electronic media, which are often relied upon so heavily to co-ordinate geographically dispersed teams (Nonaka, 1991). The project team would use electronic communication solely for transmitting project-related information such as drawings or documents to other team members. It would not be used as a forum for knowledge sharing or problem solving. E-mail messages are restricted to the printed word and are mono-directional.

As Souder (1987) suggests, new product development team members need to share perceptions and feelings, as well as factual data. When they use rich communication media, new knowledge is more likely to emerge (Nonaka, 1994). There is no evidence in either project that e-mails were used extensively to replace face-to-face dialogue. Attachments to e-mail messages enabled design inputs to be incorporated, but these were difficult to implement and were not freely used. The principal advantage of e-mail was the ability to send information to several team members at the same time and quickly. It was seldom used for social and informal discussion. Electronic messages are not as rich as those personally delivered, thus increasing the risk of misunderstanding (Canney-Davison, 1994).

Generally, it appeared that face-to-face communication in this project enhanced the understanding of pertinent issues and problems. Visual aids such as free-hand sketches, or drawings, photographs and pictures from reference books, were used to facilitate dialogue and aid further comprehension of concepts. Both project team members tended to use objects to promote the visualisation of issues. In addition, knowledge sharing through discussion embodies more of the 'human moment' than writing. Team members evidently valued interpersonal discussion and anecdotal exchange in the promotion of knowledge and social interaction.

Both project groups held formal and informal meetings on a regular basis. Formal meetings focused more on broader issues affecting all team members, such as setting goals and deadlines. Informal meetings enabled interrelated team members to resolve design issues that were affecting each other, speaking openly on issues of mutual interest. The vigorous nature of the meetings, and the detailed discussions that flowed, gave rise to a more thorough comprehension of other people's opinions. Such conditions enabled tacitly held knowledge to be converted into explicit knowledge, where team members explicitly shared previously gained experiences in order to facilitate work on the current project.

The sharing of knowledge happened naturally, as part of the discussions. Team members shared the knowledge they had gained from concurrent or previous projects. Some were able to share project knowledge, not from their own domain but from the organisational memory embedded within their companies. Others were able to offer knowledge acquired through inspecting other facilities. Some of this knowledge was acquired at weekends or on vacations overseas. The shared knowledge came from a range of sources and experiences. Communication was the key to sharing knowledge in this project team.

12.3.3   *Knowledge sharing from different knowledge domains*

Through training and experience, team members may acquire information and knowledge that others do not possess. Knowledge sharing within multi-disciplinary project teams can reveal the diverse material possessed by different professional team members.

Personal discussions, at work or during social activities, with other project team members were used extensively for knowledge sharing, to assist the problem-solving and decision-making processes. Socialisation is a valuable mode of creating knowledge within organisations (Nonaka, 1994). It enables individual team members to understand each other and work together towards common goals but from different perspectives (Saint-Onge, 1996). Team members have diverse backgrounds, training, expertise and experience. They contribute their different histories, skills and knowledge to the project, all uniquely fashioned through years of individually distinct experience and training. Team members appear to be comfortable in sharing information and knowledge, and are not at all reluctant to offer their insights to colleagues. An example in this case was that team members openly shared their opinions about the clubhouse features, based on insights from projects they had worked on before, the latest sales brochures, or new trends both in Hong Kong and overseas. Besides sharing positive experiences, team members in this project candidly shared negative experiences gained from previous projects, so that they could be avoided this time. They all agreed that even after many years, they still stumbled upon things they did not know. They seemed to perceive each project as a new learning experience.

The finding here matched the results of research by Stasser and Titus (1987), who found that members of a diverse group, each with different information to give, may be more likely to discuss uniquely held information than groups comprising similar, like-minded members (Wittenbaum and Stasser, 1996). People from diverse groups are less likely to have information in common, and will therefore have that much more to exchange. The advantage of heterogeneous teams stems from the diverse pool of accessible information and knowledge that can be shared in meetings or discussions.

The following section highlights several contributory factors identified from the research as influencing the knowledge-sharing process.

12.3.4 *Contributory factors influencing the knowledge-sharing process*

The knowledge-sharing process appears to be moulded by four different influences, namely openness, motivation, trust and pressure of time.

Openness

The openness seemed particularly evident in team meetings, with open discussion taking place concerning issues and problems. Such openness allowed team members to voice potentially useful ideas without fear of ridicule. Team members explored the issue of 'feng shui' in the luxury flats, fearing that this could affect the layout design and, ultimately, influence the decisions of potential buyers. Through knowledge they had acquired from past projects, or through reading design reference materials, their ideas were pooled to enable them to consider possible alternatives before a final decision was made. Lane and Bachmann (1998) found that openness between partners positively influences the transfer of knowledge. Since some team members on this project had previously worked with other team members, the positive experience of working with colleagues elsewhere further enhanced openness (Lane and Bachmann, 1998).

It was apparent that team members were open enough to share knowledge and experience with other professional disciplines, observed during previous or concurrent projects. In this way knowledge was shared across professional boundaries. It was seen as an opportunity to listen to opinions or solutions gained from other projects through the participation of current team members.

Motivation

Team members working on this project found that sharing knowledge provided personal satisfaction and gratification. Much time and effort could be expended searching through professional journals and manuals, and this could often be circumvented through the knowledge-sharing process. In tight design programmes, parties could ill-afford to waste time. Knowledge collaboration frequently precluded this, firmly motivating participants to share expertise and ideas.

Trust

Another influence identified was trust. Team members did not hesitate to share their knowledge within the project team, given the client organisation's open culture and the frequent efforts of project managers to allay any anxiety that could arise through the sharing of new ideas. Nam and Tatum (1992) suggested that without any contractual obligation between professionals, respect and trust appear to be strong motivators of co-operation.

Trust refers to a belief in people's capability (Szulanski, 1996), or to 'competence trust' (Newell and Swan, 2000), which is a belief in people's competence. Team members on the present project may have derived a level of trust based on the client's reputation for recruiting the best consulting firms, constantly evaluating the professional competence of all its listed consultants. This trust was further enhanced by their mutual work experience, which tended to confirm that the other team members were highly capable.

When people trust each other, they also help one another because they feel it is morally right (Tyler and Kramer, 1996). Team members here appeared willing to engage in exchanging knowledge in a co-operative manner, possibly reassured by the sentiment of trust.

In practice, the time needed to develop trust in project-based consulting might be too long, as participants often have temporary status. They probably find that they have too little time to engage in team-building or trust-building activities. Therefore, many temporary systems act as if trust were present, even though their histories seemed to preclude its development (Meyerson et al., 1996). Meyerson et al. (1996) call this phenomenon 'swift trust'. To transfer the individual expertise of strangers into interdependent work, people must reduce their uncertainty of one another through activities that resemble trust. This could have been reflected in this project, as the interviewed team members did not highlight trust as such an important issue in knowledge-based work, despite the numerous literature claims. This could suggest that trust had developed swiftly, or that people had taken a 'leap of faith', or that they simply presumed that performance needs were imperative. Since some team members had previous collective work experience, Sherif and Sherif (1953) referred to this as the benefit of building up mutual trust.

It was revealed that the establishment of trust or friendship in a project would guarantee personal ties for the future. It was felt that the relationships could extend beyond the current project, into people's informal personal networks.

Time pressure

Time pressure can act as a double-edged sword in the process of knowledge sharing. Time pressure on this project was due to insufficient human resources, team members' commitment to multiple projects, and the client's tight schedule, as well as to the corporate focus on profit maximisation. On the one hand, time constraints can restrict the opportunities for sharing knowledge (Starbuck, 1992), with participants rushing to get their work done. Members can quite easily resort to adopting previously workable solutions if time is really pressing. However, construction design should require innovative and freshly creative input, in order to attract customers and improve previous products.

On the other hand, team members recognised that deadlines could stimulate knowledge sharing so that time could be effectively managed, with the risks of error possibly diminished. It was evident in this project that the team members did share their knowledge by lending previous drawings or designs to other team members for reference. This often helped reduce the design time by minimising preliminary investigations and highlighting any past imperfections or pitfalls.

12.4   Conclusions

This study suggests that competition can be detrimental to the knowledge-sharing process. Sharing important market or design knowledge can facilitate imitation by competitors, possibly even resulting in a project being poached by another consulting firm. In addition, it proposes that the type of communication is more important in the transfer of tacit knowledge than in that of explicit knowledge. For tacit knowledge to be transmitted, interpersonal communication was of utmost importance, as team members shared tacitly held personal experiences through dialogue. Orr (1990) demonstrated how narrative, in the form of stories, facilitates the exchange of practice and tacit experience between technicians. The emergence of shared narratives within a team enables the creation and transfer of new interpretations of events facilitating the combination of different forms of knowledge, including those that are largely tacit. Marwick (2001, p. 815) found that 'through conceptualisation, elicitation, and ultimately articulation, typically in collaboration with others, some proportion of a person's tacit knowledge may be captured in explicit form'. He suggests several activities that will enable the sharing of tacit knowledge, including dialogue among team members, response to questions and story-telling.

The experiences of this multi-disciplinary project team also revealed that team members from differing knowledge domains were more likely to discuss their uniquely held information and knowledge than those who held information in common. It was an advantage to have a diverse pool of knowledge that team members could access and share in meetings or discussions. It is also clear that four influences appear to encourage the sharing of knowledge. These are openness, motivation, trust and time pressure, and they all seem to affect knowledge sharing positively as well as negatively.

References

Canney-Davison, S. (1994) Creating a high performance international team. *Journal of Management Development*, **13**(2), 577–607.

Habermas, J. (1987) *The Theory of Communicative Action: Lifeworld and Social System*. Beacon Press, Boston.

Kogut, B. and Zander, U. (1992) Knowledge of the firm, combinative capabilities, and the replication of technology. *Organization Science*, **3**(3), 383–97.

Lane, C. and Bachmann, R. (1998) *Trust Within and Between Organisations*. Oxford University Press, Oxford.

Leonard, D. and Sensiper, S. (1998) The role of tacit knowledge in group innovation. *California Management Review*, **40**(3), 112–32.

Marwick, A.D. (2001) *Knowledge management technology. IBM Systems Journal*, **40**(4), 814–30.

Meyerson, D., Weick, K.E. and Kramer, R. (1996) Swift trust and temporary groups. In: *Trust in Organizations: Frontiers of Theory and Research* (T.R. Tyler and R.M. Kramer, eds), pp. 166–95. Sage, London.

Nam, C.H. and Tatum, C.B. (1992) Non-contractual methods of integration on construction projects. *Journal of Construction Engineering and Management, ASCE*, **118**(2), 385–98.

Newell, S. and Swan, J. (2000) Trust and inter-organizational networking. *Human Relations*, **53**(10), 1287–328.

Nonaka, I. (1991) The knowledge creating company. *Harvard Business Review*, **69**(6), 96–104.

Nonaka, I. (1994) A dynamic theory of organizational knowledge creation. *Organization Science*, **5**(1), 14–37.

Nonaka, I. and Konno, N. (1998) The concept of 'Ba': building a foundation for knowledge creation. *California Management Review*, **40**(3), 40–54.

Nonaka, I. and Takeuchi, H. (1995) *The Knowledge-Creating Company: How Japanese Companies Create the Dynamics of Innovation*. Oxford University Press, Oxford.

Orr, J.E. (1990) Sharing knowledge, celebrating identity: community memory in a service culture. In: *Collective Remembering* (D. Middleton and D. Edwards, eds), pp. 169–89. Sage, London.

Orr, J.E. (1996) *Talking About Machines: An Ethnography of a Modern Job*. ILR Press, Ithaca, New York.

Saint-Onge, H. (1996) Tacit knowledge: the key to the strategic alignment of intellectual capital. *Planning Review*, **24**(2), 10–14.

Sherif, M. and Sherif, C.W. (1953) *Groups in Harmony and Tension*. Harper and Brothers, New York.

Skyrme, D. and Amidon, D.M. (1997) *Creating the Knowledge-Based Business*. Business Intelligence Ltd, London.

Souder, W.E. (1987) *Managing New Product Innovations*. Lexington Books, Lexington, Massachusetts.

Starbuck, W.H. (1992) Learning by knowledge intensive firms. *Journal of Management Studies*, **29**(6), 713–40.

Stasser, G. and Titus, W. (1987) Effects of information load and percentage of shared information on the dissemination of unshared information during group discussion. *Journal of Personality and Social Psychology*, **53**(1), 81–93.

Szulanski, G. (1996) Exploring internal stickiness: impediments to the transfer of best practice within the firm. *Strategic Management Journal*, **17** (winter special issue), 27–43.

Tyler, T.R. and Kramer, R.M. (1996) Whither trust? In: *Trust in Organizations. Frontiers of Theory and Research* (T.R. Tyler and R.M. Kramer, eds), pp. 1–16. Sage, London.

von Hippel, E. (1988) *The Sources of Innovation.* Oxford University Press, New York.

Wenger, E., McDermott, R. and Synder, W. (2002) *Cultivating Communities of Practice: A Guide to Managing Knowledge.* Harvard Business School Press, Boston.

Winter, S.G. (1987) Knowledge and competence as strategic assets. In: *The Competitive Challenge: Strategies for Industrial Innovation and Renewal* (D. Teece, ed.), pp. 159–84. Ballinger, Cambridge, Massachusetts.

Wittenbaum, G.M. and Stasser, G. (1996) Management of information in small groups. In: *What's Social About Social Cognition? Research on Socially Shared Cognition in Small Groups* (J.L. Nye and A.M. Brower, eds), pp. 3–28. Sage, Thousand Oaks, California.

13 Concluding Notes

Chimay J. Anumba, Charles O. Egbu
and Patricia M. Carrillo

13.1 Introduction

This chapter concludes the book and draws together a number of the threads running through the various chapters. It starts with a brief summary of the contents of the book, restates the main benefits of knowledge management (KM) to construction-sector organisations and highlights a number of challenges in the implementation of KM. The last section of the chapter explores some of the future directions for KM.

13.2 Summary

This book has sought to introduce the concept of KM to construction-sector organisations. This is in recognition of the growing importance of knowledge (and intellectual capital) as a major source of competitive advantage in the business world. In this respect, the nature of construction projects and the teams of people involved make it imperative that construction-sector organisations make more effective use of their knowledge. They need to recognise that KM is more than a 'buzz word'; it is at the very essence of an organisation's core competences and business processes and should not be ignored.

The first part of the book (Chapters 1–4) provided a general background to the subject and how it relates to the construction industry. Chapters 1 and 2 set the scene by defining KM, and exploring in depth the nature and dimensions of KM (from a wide, non-construction perspective). The growing interest in KM was explained and the importance of absorptive capacity highlighted. Chapter 3 focused on the place of construction as a knowledge-based industry and highlighted the key aspects of creating and sustaining a knowledge culture in a knowledge economy. The variety of knowledge that needs to be managed in construction organisations was also explored. The importance of strategies and business cases for KM was discussed in Chapter 4. Key questions that need to be answered in making the business case for KM were also outlined and details of the approach adopted by one of the leading firms in the UK construction industry presented.

The second part of the book (Chapters 5–8) addressed practical organisational aspects of KM. The extent to which an organisation is ready for the implementation of KM was covered in Chapter 5 and it was emphasised that organisations need to take steps to ensure that they are adequately prepared before putting KM initiatives in place. Chapter 6 discussed the tools and techniques that organisations may adopt in the implementation of KM. The importance of selecting the most appropriate tools and techniques in a given situation was highlighted and a structured framework (SeLEKT) presented, which enables organisations to make an informed selection based on the knowledge dimensions of interest and their business needs and context. Given the fact that construction is a project-based industry, Chapter 7 explored the issues involved in cross-project KM. Issues covered included the nature of construction projects (and teams), the characteristics of knowledge in a multi-project environment, and the role of individuals, project reviews, and contractual and organisational arrangements on knowledge transfer. The live capture of knowledge in projects was advocated as the way forward and a conceptual framework for how this might be implemented was outlined. The role of KM as a driver for innovation in an organisation was discussed in Chapter 8. It was argued that an organisation's ability to innovate, and its profitability, depend to a large extent on how well it manages its knowledge. The implications for construction managers were also identified.

The third part of the book (Chapters 9–12) was dedicated to addressing some of the difficult and often knotty issues involved in managing knowledge in an organisation. The first of these, performance measurement, was the subject of Chapter 9. The chapter emphasised the need to measure performance and explored the difficulties faced by many organisations in measuring KM performance. Existing approaches were reviewed and their advantages and limitations summarised. Chapter 10 was concerned with the development of a KM strategy as it is often the most critical aspect of KM implementation. It presented the CLEVER framework, which enables an organisation to ensure that its KM strategy is directly related to its business objectives and addresses a grounded rather than a perceived problem. Chapter 11 focused on the issues associated with establishing a corporate memory, based on which knowledge capture, sharing and reuse can be undertaken. It presented a number of innovative and interactive IT tools that support KM using a corporate memory. Another difficult area, how to build a culture of knowledge sharing in construction project teams, was covered in Chapter 12. The importance of socialising, sharing negative as well as positive experiences, openness, trust, motivation and time as essential ingredients in knowledge sharing were highlighted and a case study used to illustrate the key principles.

13.3   Benefits of knowledge management to construction organisations

Many of the chapters in this book have either explicitly stated the benefits of KM to construction organisations or alluded to them. In making the decision on whether or not to adopt KM, construction organisations need to recognise that the costs of not managing knowledge effectively are as important as the benefits. However, it is necessary to reiterate the potential benefits of improved KM:

- Innovation is more likely to thrive in an environment where there is a clear strategy for managing knowledge.
- Improved performance will result from the pooling of an organisation's knowledge as workers will be both more effective (adopting the most appropriate solutions) and more efficient (using less time and other resources).
- KM is vital for improved construction project delivery, as lessons learned from one project can be carried on to future projects, resulting in continuous improvement.
- KM can facilitate the transfer of knowledge across a variety of project interfaces (participants, disciplines, organisations, stages, etc.).
- With effective KM, firms and project teams can avoid repeating past mistakes and/or reinventing the wheel.
- Increased intellectual capital is a major benefit for many organisations, which is able to narrow the gap between what employees know and what the organisation knows.
- Firms that adequately manage their knowledge are better placed to respond quickly to clients' needs and other external factors.
- KM results in improved support for teams of knowledge workers in an organisation or project team.
- Dissemination of best practice is one of the results of knowledge sharing – this can happen both within and across organisations.
- Organisations can retain the tacit knowledge that would otherwise be lost when valued employees leave, retire or die.
- Increased value can be provided to the customers of construction organisations through better management of knowledge.
- With effective KM, construction organisations can be more agile and better able to respond to organisational changes.
- Risk minimisation is one of the key benefits of KM, as the enhanced knowledge base means that organisations have fewer uncertainties to deal with.

13.4   Issues in knowledge management implementation

Construction organisations intending to adopt KM need to address a number of key issues to ensure that they maximise the benefits outlined above. It is particularly important that an organisation does not implement KM simply because other organisations are doing so. Instead, the implementation should be backed by a well-thought-out KM strategy that derives from the organisation's business objectives. Some of the other considerations in KM implementation include the following:

- Establishing what the KM problem is prior to making any investments in KM tools and techniques.
- The characteristics of the knowledge (e.g. tacit or explicit) that the organisation is interested in managing need to be established, as these have huge implications regarding the approach to be adopted.
- It is important to identify the location of the knowledge to be managed and any associated availability and/or access constraints.
- In some situations, an organisation may need to know how the knowledge is acquired. For example, is it best acquired through a formal course of instruction or can it be acquired simply by informal interaction with people that have the knowledge?
- Prior to implementing KM, organisations need to assess their culture, as this is considered to be a far greater influence on the success of KM initiatives than purely technical issues.
- It is important to identify and secure 'buy-in' from all potential stakeholders likely to be impacted by any KM initiative; this will include managers, knowledge providers and knowledge users.
- KM is best implemented in 'bite-size chunks'; this means that it is inadvisable to seek to roll out KM across the whole organisation in one go. It is important to select a business unit, geographical area or business process within which to initiate KM.
- There should be clear metrics for evaluating the effectiveness of any KM implementation; these should allow for periodic monitoring so that changes can be made as appropriate.

KM implementation at construction project level is complicated by the transient nature of 'virtual' project organisations, which often involve different organisational priorities during a project. It is important that long-term partnerships are established within which an effective knowledge-sharing culture can thrive.

13.5   Future directions

KM existed long before the phrase came into common parlance. Construction organisations, in line with other organisations, have always had mechanisms for managing the knowledge that they possess, even when they did not regard themselves as practising KM. The current focus on KM has forced many organisations to review some of these mechanisms and to improve their effectiveness. Thus, in exploring the future directions of KM, it is important to recognise that KM practices will continue long after the phrase has waned in popularity. It is impossible to predict with any accuracy the direction that KM will take in the future, so this section will simply outline a number of developments that will have an impact on shaping the future:

- KM will increasingly become an integral part of an organisation's business processes in much the same way that computer-aided engineering (CAE) systems are now well integrated into the operations of most engineering firms.
- One of the biggest issues in KM implementation is performance assessment. This is because knowledge managers are increasingly required to justify the investment in KM. No universally accepted metrics exist and it will become increasingly necessary for organisations to develop measures that are tailored to their circumstances; these will have to be linked to higher-level corporate objectives.
- Developments in intelligent agent technology will result in a growth in agent-based KM systems. Such systems will be able to act on behalf of their owners in interacting with any existing IT tools and knowledge repositories, and can build user profiles based on a course of interaction.
- It is expected that more sophisticated tools and techniques for knowledge capture and reuse will become available. Some of these will be based on artificial intelligence (AI) techniques and enable more context-specific and, where appropriate, location-based services.
- It is becoming increasingly recognised that the market value of an organisation has to be measured not only by its tangible assets but also by its intangible assets – knowledge and intellectual property.
- There is likely to be an increase in the number of people whose roles involve the management of knowledge assets in organisations. Education and training provisions for these individuals are also likely to increase.

Clearly, KM has much to offer construction-sector organisations. The complexity associated with the delivery of construction projects by a transient project team made up of individuals/teams from a variety of organisations makes the implementation of KM challenging. However, it is also for the same reasons that construction-sector organisations cannot afford to ignore KM. They must imbibe the core principles and tailor their KM strategy and implementation to the vagaries of their particular firms. At project level, the virtual organisation also needs to embrace KM so as to facilitate 'live' capture of knowledge and collaborative learning during the course of the project. For those firms and project teams that get it right, the rewards will far outweigh the investment and ensure that they rise above the competition.

Index